# DECLINE OF THE STELLER SEA LION IN ALASKAN WATERS

## UNTANGLING FOOD WEBS AND FISHING NETS

Committee on the Alaska Groundfish Fishery and Steller Sea Lions

Ocean Studies Board
Polar Research Board

Division on Earth and Life Studies
NATIONAL RESEARCH COUNCIL
*OF THE NATIONAL ACADEMIES*



The National Academies Press
Washington, D.C.
**www.nap.edu**

**THE NATIONAL ACADEMIES PRESS 500 Fifth Street, N.W. Washington, DC 20001**

NOTICE: The project that is the subject of this report was approved by the Governing Board of the National Research Council, whose members are drawn from the councils of the National Academy of Sciences, the National Academy of Engineering, and the Institute of Medicine. The members of the committee responsible for the report were chosen for their special competences and with regard for appropriate balance.

This study was supported by Contract/Grant No. SSL-01-B between the National Academy of Sciences and the North Pacific Fishery Management Council. Any opinions, findings, conclusions, or recommendations expressed in this publication are those of the author(s) and do not necessarily reflect the views of the organizations or agencies that provided support for the project.

Library of Congress Control Number: 2003101402
International Standard Book Number: 0-309-08632-9

Cover art was created by Emma Wilson. Ms. Wilson is the founder and principal of the Emma Wilson Design Company in Seattle, Washington. She has worked with many Seattle area companies on the design of annual reports, promotions, posters, corporate and retail branding, and environmental graphics.

Additional copies of this report are available from the National Academies Press, 500 Fifth Street, N.W., Lockbox 285, Washington, DC 20055; (800) 624-6242 or (202) 334-3313 (in the Washington metropolitan area); Internet, http://www.nap.edu



Printed in the United States of America

The **National Academy of Sciences** is a private, nonprofit, self-perpetuating society of distinguished scholars engaged in scientific and engineering research, dedicated to the furtherance of science and technology and to their use for the general welfare. Upon the authority of the charter granted to it by the Congress in 1863, the Academy has a mandate that requires it to advise the federal government on scientific and technical matters. Dr. Bruce M. Alberts is president of the National Academy of Sciences.

The **National Academy of Engineering** was established in 1964, under the charter of the National Academy of Sciences, as a parallel organization of outstanding engineers. It is autonomous in its administration and in the selection of its members, sharing with the National Academy of Sciences the responsibility for advising the federal government. The National Academy of Engineering also sponsors engineering programs aimed at meeting national needs, encourages education and research, and recognizes the superior achievements of engineers. Dr. Wm. A. Wulf is president of the National Academy of Engineering.

The **Institute of Medicine** was established in 1970 by the National Academy of Sciences to secure the services of eminent members of appropriate professions in the examination of policy matters pertaining to the health of the public. The Institute acts under the responsibility given to the National Academy of Sciences by its congressional charter to be an adviser to the federal government and, upon its own initiative, to identify issues of medical care, research, and education. Dr. Harvey V. Fineberg is president of the Institute of Medicine.

The **National Research Council** was organized by the National Academy of Sciences in 1916 to associate the broad community of science and technology with the Academy's purposes of furthering knowledge and advising the federal government. Functioning in accordance with general policies determined by the Academy, the Council has become the principal operating agency of both the National Academy of Sciences and the National Academy of Engineering in providing services to the government, the public, and the scientific and engineering communities. The Council is administered jointly by both Academies and the Institute of Medicine. Dr. Bruce M. Alberts and Dr. Wm. A. Wulf are chair and vice chair, respectively, of the National Research Council.

**www.national-academies.org**

## COMMITTEE ON THE ALASKA GROUNDFISH FISHERY AND STELLER SEA LIONS

**ROBERT T. PAINE** (*Chair*), University of Washington, Seattle
**DANIEL W. BROMLEY**, University of Wisconsin, Madison
**MICHAEL A. CASTELLINI**, University of Alaska, Fairbanks
**LARRY B. CROWDER**, Duke University, Beaufort, North Carolina
**JAMES A. ESTES**, U.S. Geological Survey/University of California, Santa Cruz
**JACQUELINE M. GREBMEIER**, University of Tennessee, Knoxville
**FRANCES M.D. GULLAND**, The Marine Mammal Center, Sausalito, California
**GORDON H. KRUSE**, University of Alaska, Fairbanks
**NATHAN J. MANTUA**, University of Washington, Seattle
**JAMES D. SCHUMACHER (a.k.a. Two Crow)**, Two Crow Environmental, Inc., Silver City, New Mexico
**DONALD B. SINIFF**, University of Minnesota, St. Paul
**CARL J. WALTERS**, University of British Columbia, Vancouver

**Staff**

**SUSAN J. ROBERTS**, Study Director
**NANCY A. CAPUTO**, Senior Project Assistant

The work of this committee was overseen by the Ocean Studies Board and the Polar Research Board of the National Research Council (See Appendix G).

# Preface

The National Research Council (NRC) has often been called on to provide unbiased and nonjudgmental evaluation of issues that are simultaneously significant and vexing. The Steller sea lion committee was assembled to address just such a challenge. Populations of these sea lions, especially in the western Gulf of Alaska and Bering Sea, have been in decline since about 1970. They were listed as endangered in 1997 under the Endangered Species Act. This enormous region is also the site of the nation's most valuable groundfish industry, yielding a dockside value approaching $700 million in 2000. Since sea lions eat groundfish, it is not illogical to suspect a causal relationship between their decline and industry removals of about 4 billion pounds of their potential prey per year. This possibility was explored in earlier analyses, including the first "Is It Food?" conference, which concluded that food availability likely contributed to the decline.

Data gathered since 1990 suggest that alternative hypotheses deserve equal scrutiny, and the sea lion committee undertook this task. It is not that these hypotheses are new or unanticipated; nor has the older food limitation hypothesis been ignored. Instead, the committee has tried to sift through the enormous but still frustratingly limited database relevant to resolving the question of why these sea lions continue to decline despite the imposition of substantial constraints on fishers. In the past decade, some analyses supported the ecologically plausible possibility of food limitation (a bottom-up control hypothesis), but this committee's numerous consultations, evaluation of recent data, and modest modeling efforts suggest that mortality imposed on sea lions by their enemies, including humans (top-down hypotheses), could also constitute elusive but convincing explanations.

Both viewpoints embody generalizations: both the bottom-up and top-down perspectives can be addressed by scientific evidence. Emotion, taxonomic advocacy, and regulatory roadblocks to legitimate research have constrained a science-based understanding of why sea lions continue to decline. Good science may not be able to save sea lions from regional extinction, but it is certain that without the understanding that focused research provides, the causes of the decline can neither be understood nor addressed.

So with much of the above understood a priori, the committee chose to address the issue of data gaps, historical trends in the mass of fishes caught and their species composition, the intimate details of sea lion biology, and to offer a collective opinion on what actions should be taken next. That the committee has done, although the advice may please neither fishing industry nor sea lion advocates. The committee met four times: twice in Seattle where the National Marine Fisheries Service's Alaska Fisheries Science Center and many of the relevant scientific experts are located, once in Anchorage, largely to gather information from interested and involved nongovernmental organizations and Alaska natives, and once in Florida to complete the writing of the report.

I know that all scientists willing to join committees like this one lead busy and complex lives. The same must surely hold true for the NRC staffers and presenters at our three public committee meetings. The organizational challenge, which must be like the proverbial "herding cats," was superbly managed by study director Susan Roberts and project assistant Nancy Caputo. On behalf of the committee, I thank them. It is also appropriate here to acknowledge the willingness of the committee's members to work toward the common goal of meeting the statement of task despite their disparate backgrounds and opinions. It is a rare privilege to listen to experts debate issues as complex as this one characterized by substantial data gaps. It is even more gratifying to participate in an effort where numerous potential pitfalls and strongly polarized opinions have been minimized. The committee strove to develop a constructive analysis of a continuingly contentious issue. I believe we succeeded.

Robert T. Paine, *Chair*
Committee on the Alaska Groundfish
Fishery and Steller Sea Lions

# Acknowledgments

This report was greatly enhanced by the contributions of many people. We would like to thank the sponsor of this study, the North Pacific Fishery Management Council, executive director, Chris Oliver, and staffers, David Witherell, Jane DiCosimo, and Gail Bendixen, who provided valuable assistance and information. We also thank Thomas Loughlin and Anne York of the National Marine Mammal Laboratory, Alaska Fisheries Science Center, for providing data and figures for this report. John Sease, Lowell Fritz, and Patsy Bearden of the National Oceanic and Atmospheric Administration kindly sent us electronic figures that are reproduced here. Alison Rieser of the Marine Law Institute, University of Maine School of Law, Portland, Maine, provided the committee with valuable information on marine law and policy.

The committee also wishes to acknowledge the efforts of those who gave presentations at one of the three public committee meetings. These presentations gave the committee up-to-date information on research relevant to the issues addressed in this study.

**D. Lee Alverson**, Natural Resources Consultants, Inc., Seattle

**Russel Andrews**, Institute of Marine Science, University of Alaska, Fairbanks, and the Alaska SeaLife Center, Seward

**Kerim Aydin**, Alaska Fisheries Science Center, National Marine Fisheries Service, Seattle

**Vernon Byrd**, Alaska Maritime National Wildlife Refuge, U.S. Fish and Wildlife Service, Homer

**Vladimir Burkanov**, Natural Resources Consultants, Inc., Seattle
**Shane Capron**, National Marine Fisheries Service, Juneau
**Helen Chythlook**, Bristol Bay Native Association, Dillingham, Alaska
**David Cline**, retired, formerly with World Wildlife Fund, Anchorage
**Larry Cotter**, Aleutian Pribilof Island Community Development Association, Juneau
**Douglas DeMaster**, Alaska Fisheries Science Center, National Marine Fisheries Service, Seattle
**Douglas Eggers**, Alaska Department of Fish and Game, Juneau
**David Fraser**, owner and captain, *F/V Muir Milach*, Port Townsend, Washington
**Lowell Fritz**, Alaska Fisheries Science Center, National Marine Fisheries Service, Seattle
**Fritz Funk**, Alaska Department of Fish and Game, Juneau
**Thomas Gelatt**, Alaska Department of Fish and Game, Anchorage
**Daniel Goodman**, Montana State University, Bozeman
**Karl Haflinger**, Sea State, Inc., Vashon, Washington
**Oliver Holm**, fisherman, Kodiak, Alaska
**George Hunt, Jr.**, School of Biological Sciences, University of California, Irvine
**Linda Larson**, Sandler, Ahern & McConaughy, Seattle
**Frank Logusak**, Togiak Traditional Council, Togiak, Alaska
**Thomas Loughlin**, National Marine Mammal Laboratory, Alaska Fisheries Science Center, National Marine Fisheries Service, Seattle
**Herbert Maschner**, Department of Anthropology, Idaho State University, Pocatello
**Chris Oliver**, North Pacific Fishery Management Council, Anchorage
**Kenneth Pitcher**, Alaska Department of Fish and Game, Anchorage
**Timothy Ragen**, Marine Mammal Commission, Bethesda, Maryland
**Alan Springer**, Institute of Marine Science, University of Alaska, Fairbanks
**Jack Sterne**, Trustees for Alaska, Anchorage
**Kenneth Stump**, consultant for Greenpeace, Seattle
**Jack Tagart**, Washington Department of Fish and Wildlife, Olympia, and the North Pacific Fishery Management Council, Scientific and Statistical Committee
**Gary Thomas**, Prince William Sound Science Center, Cordova, Alaska
**Andrew Trites**, Fisheries Centre, University of British Columbia, Vancouver
**Glenn VanBlaricom**, School of Aquatic & Fisheries Sciences, University of Washington, Seattle

**Terrie Williams**, Center for Ocean Health, University of California, Santa Cruz

**David Witherell**, North Pacific Fishery Management Council, Anchorage

**Boris Worm**, Biology Department, Dalhousie University, Halifax, Canada

**Kate Wynne**, Sea Grant Marine Advisory Program, University of Alaska, Kodiak

**Anne York**, National Marine Mammal Laboratory, Alaska Fisheries Science Center, National Marine Fisheries Service, Seattle

This report has been reviewed in draft form by individuals chosen for their diverse perspectives and technical expertise, in accordance with procedures approved by the National Research Council's (NRC) Report Review Committee. The purpose of this independent review is to provide candid and critical comments that will assist the institution in making its published report as sound as possible and to ensure that the report meets institutional standards for objectivity, evidence, and responsiveness to the study charge. The review comments and draft manuscript remain confidential to protect the integrity of the deliberative process. We wish to thank the following individuals for their participation in the review of this report:

**D. Lee Alverson**, Natural Resource Consultants, Inc., Seattle

**David R. Cline**, retired, formerly with the World Wildlife Fund, Anchorage

**Richard B. Deriso**, Inter-American Tropical Tuna Commission, La Jolla, California

**Thomas Gelatt**, Alaska Department of Fish and Game, Anchorage

**Daniel Goodman**, Montana State University, Bozeman, Montana

**Susan M. Henrichs**, Institute of Marine Science, University of Alaska, Fairbanks

**Henry P. Huntington**, Huntington Consulting, Eagle River, Alaska

**Andrew A. Rosenberg**, University of New Hampshire, Durham

**Douglas Wartzok**, Florida International University, Miami

Although the reviewers listed above provided many constructive comments and suggestions, they were not asked to endorse the conclusions or recommendations nor did they see the final draft of the report before its release. The review of this report was overseen by May R. Berenbaum, University of Illinois, Urbana, Illinois, and by Kenneth Brink,

Woods Hole Oceanographic Institution, Woods Hole, Massachusetts. Appointed by the NRC, they were responsible for making certain that an independent examination of this report was carried out in accordance with institutional procedures and that all review comments were carefully considered. Responsibility for the final content of this report rests entirely with the authoring committee and the institution.

# Contents

# Executive Summary

Theory helps us bear our ignorance of fact.
—George Santayana, *The Sense of Beauty*

Steller sea lions are found along the North Pacific rim from California to Japan with about 70% of the population living in Alaskan waters. The Alaskan population declined precipitously during the 1970s and 1980s and continued to decline at a slower rate during the 1990s. Overall, the Alaskan population has declined by more than 80% over the past 30 years. In 1990 the Steller sea lion was listed as a threatened species, and in 1997 the population west of Cape Suckling (longitude 144° W) was listed as endangered under the Endangered Species Act (ESA). The eastern population (southeast Alaska to California) has increased gradually throughout most of its range since the 1970s, but this stock remains listed as threatened.

The causes of the decline of the western stock have been the subject of much speculation and debate despite numerous analyses and many detailed reports. There is no widely accepted answer to the question of why the Steller sea lion population is declining. What might otherwise be an obscure ecological mystery has become an issue of great regional and even national interest because of the regulatory implications for management of the large commercial fisheries in the North Pacific. These fisheries target many of the fish species that comprise the prey base for Steller sea lions.

**BOX ES.1**
**Statement of Task**

This study will examine interactions between Alaska groundfish fisheries and Steller sea lions (*Eumetopias jubatus*) and the role of these fisheries in the evolving status of the sea lion population. The focus of the study will be (1) the status of current knowledge about the decline of the Steller sea lion population in the Bering Sea and Gulf of Alaska ecosystems; (2) the relative importance of food competition and other possible causes of population decline and impediments to recovery; (3) The critical information gaps in understanding the interactions between Steller sea lions and Alaska fisheries; (4) the type of research programs needed to identify and assess potential human and natural causes of sea lion decline; and (5) the components of an effective monitoring program, with yardsticks for evaluating the efficacy of various management approaches.

In November 2000 the ESA consultation prepared by the National Marine Fisheries Service concluded that the Alaska groundfish fishery posed a threat to the recovery of the Steller sea lion and imposed more restrictive measures on the management of the fishery. Concern that the new regulations would bring significant social and economic disruption prompted Congress to direct the North Pacific Fishery Management Council to sponsor an independent scientific review by the National Academy of Sciences on the causes of the Steller sea lion decline and the potential efficacy of the new management measures (Box ES.1). This report represents the results of that review.

## CAUSES OF DECLINE

Over the past 200 years many populations of terrestrial and marine mammals have declined precipitously, some to the point of extinction. Most declines of marine mammals have been attributed to human activities, typically as a result of commercial harvest for fur, meat, and oil or because of fishery interactions, through incidental catch in fishing nets, disturbance from fishing activities, or predator control programs. Suspension of these activities reduces the risk of extinction, but for some long-lived species recovery may take decades.

The case of the dramatic decline in the Steller sea lion population has been less straightforward. Steller sea lions have not been subject to large commercial harvests since 1972, and the take of sea lions by fisheries has been estimated to be small relative to the size of the population. During the period of rapid population decrease during the late 1970s through the

1980s, there were also major shifts in abundance of many marine species in the North Pacific attributed to both climatological events and commercial harvests of fishes. Analysis of these trends has been complicated by the scarcity of baseline population data on the robust sea lion population that existed before 1975, which is needed for comparison with data on the current depleted population. Since there are few avenues for augmenting this historical database (e.g., reanalysis of existing data, testing of archived tissue samples for contaminants and disease agents, reconstruction of environmental events based on isotope anomalies or annual growth patterns), the cause, or causes, of the early phase of the sea lion population decline will likely remain a source of speculation and debate. However, existing information can be used to identify scenarios that could explain the historical decline, which will be valuable in understanding the prospects for recovery of the remaining population.

Under the ESA, federal agencies must ensure that their actions, or actions they authorize, are not likely to jeopardize the survival or recovery of protected species or damage the protected species' critical habitat. Therefore, if a federally regulated activity may affect Steller sea lions, the responsible agency must take actions to ensure that negative impacts are avoided. This requirement has made it imperative to identify human activities that may contribute to the decline of Steller sea lions so that regulatory actions can be adjusted to address threats to the western population's survival. Unlike the biological opinions required by the ESA listing, this report does not assess the statutory basis for regulating the groundfish fisheries.

At least eight plausible hypotheses have been proposed to explain the decline of the sea lion population. These include threats that result from human activities and naturally occurring events that affect sea lion survival. Human activities that may threaten sea lion recovery include direct takes such as illegal shooting and subsistence harvest, and incidental takes through capture or entanglement in fishing gear. Indirectly, commercial fisheries may disrupt feeding patterns, breeding, and other aspects of sea lion behavior. Also, fishing may decrease the carrying capacity of sea lion habitat through the removal of prey species or by shifting the distribution of species such that less nutritious fish dominate the prey base, the so-called junk food hypothesis. Pollution may pose another indirect effect by impairing the health of sea lions and increasing their susceptibility to disease.

But increased mortality of sea lions may not be just a consequence of human activities. There are natural cycles of abundance and decline in marine ecosystems that are driven by climate variability, predator-prey interactions, and invasions by infectious diseases or toxic algal blooms. It is difficult, and often impossible, to resolve the relative contributions of human and natural sources of change, especially since complex interactions among species may cause the combined effects to be significantly different from the effects of any single factor.

In part because of the absence of definitive data confirming or excluding any particular hypothesized cause of decline, the regulatory measures taken in response to the protected status of the western population under the ESA have been particularly contentious. **Resolution of this conflict requires management that not only improves chances for the recovery of Steller sea lions but also facilitates scientific study of the efficacy of these protective measures.**

## MAKING THE MOST OF EXISTING INFORMATION

The hypotheses proposed to explain the decline of the western stock fall into two categories. The first category, the bottom-up hypotheses, includes potential causes that would affect the physical condition of sea lions such as

- large-scale fishery removals that reduce the availability or quality of prey species,
- a climate/regime shift in the late 1970s that changed the abundance or distribution of prey species,
- nonlethal disease that reduced the foraging efficiency of sea lions, and
- pollutants concentrated through the food web that contaminated fish eaten by sea lions, possibly reducing their fecundity or increasing mortality.

The second category, the top-down hypotheses, encompasses factors that kill sea lions independently of the capacity of the environment to support the sea lion population. These include

- predators such as killer whales (or possibly sharks) that switched their prey preference to sea lions,
- incidental takes of sea lions through capture or entanglement in fishing gear that increased as a result of the expansion of commercial fisheries,
- takes of sea lions in the subsistence harvest that were higher than estimated,
- shootings of sea lions that were underestimated in the past and present, and
- pollution or disease that increased mortality independently of effects on nutrition (e.g., introduction of a contagious pathogen could

decimate a population and give the same appearance as an efficient predator).

Observed characteristics of sea lion biology and behavior should be different under these two categories. The bottom-up hypotheses predict increased mortality through reduction in physical condition, manifested by changes in physiology, reproductive success, and foraging behavior. Top-down hypotheses predict no loss of individual fitness but require increased activity by predators, people, or pathogens. Hence, indicators of sea lion health and feeding behavior may be informative in distinguishing the likelihood of these two modes of sustained population decline. It is important to remember that some combination of both types of factors may have contributed to the population decline. For instance, evidence indicating a significant decrease in sea lion physical condition would not exclude the possibility that top-down causes also contributed to overall mortality. Also, geographic variations in environmental conditions across the range of the western population may mean that different factors are to varying degrees responsible for mortality in different parts of the range.

In the existing body of information on Steller sea lions, there is no conclusive evidence supporting either the bottom-up or the top-down hypotheses. Therefore, the available data must be carefully evaluated to ascertain the more plausible causes. First, the evidence can be categorized according to the time period during which it was collected. The rate of decline of the western population has changed since it began in the 1970s. From 1975 to 1985, the annual rate of decline averaged 5.9%. Over the next 5 years the population dropped precipitously, about 15.6% per year. Since the early 1990s (through 2001), the population has continued to decrease but at the more gradual rate of 5.2% annually. The loss of such a large fraction of the population during a relatively short time span (1985-1990) indicates that sea lions were subject to a threat, or threats, that spurred the decline in the 1980s but that by the 1990s these threats either had ended or had less impact.

Second, the evidence can be sorted geographically. In 1995 the National Oceanic and Atmospheric Administration determined that Steller sea lions west of 144° W constituted a distinct population unit based on dispersal patterns, population trends, and genetic differentiation. Because female Steller sea lions tend to return to their natal rookeries for breeding, the western stock may be considered a metapopulation. A metapopulation is a regional population comprised of semi-isolated local populations with limited exchange or interaction, which may fluctuate in response to regional as well as global impacts. Hence, variability in the geographic pattern of decline may point to causes that are specific to particular areas.

Temporal and spatial evaluation of the population data show that the 5-year period of rapid decline (1985-1989) was a range-wide phenomenon and hence was most likely caused by an ecosystem-wide change in the Steller sea lion's environment. Hypotheses that are consistent with this pattern include nutritional limitation through competition with fisheries and changes in prey abundance due to the environmental regime shift in the late 1970s, predators switching from a depleted prey population to sea lions, or introduction of a lethal and highly contagious disease agent such as a virus. Evidence for nutritional limitation includes observations that sea lion condition, growth, and reproductive performance were lower during this time period. However, ecosystem models based on data from the eastern Bering Sea indicate that changes in the relative abundance of prey cannot account for the full magnitude of the decline. Either increased predation or epidemic disease could account for the high mortality rate, but systematic observations of killer whale (or possibly shark) predation were not conducted at that time and serological tests to date have been negative for common pathogens associated with disease epidemics in marine mammals. The large increase in the rate of decline was unlikely to be caused primarily by subsistence harvest, toxic algal blooms, or illegal shooting because these threats tend to vary by geographic location and there is no evidence to suggest that they greatly intensified during this time period. **Multiple factors probably contributed to the widespread population decline in the 1980s, including incidental and deliberate mortality associated with fishing activities, but elucidation of the complete spectrum of causes and consequences is unlikely because of gaps in the available data**.

The pattern of decline has changed since the early 1990s. Not only has the overall rate of decline decreased, but individual rookeries show different population trends as well. Over the past decade, the majority have continued to decline, some have stayed at the same level, and a few have shown modest increases. Based on the most recent census of trend sites, counts of adults and juveniles in 2002 show a 13.6% increase in the Gulf of Alaska and less than a 1% decrease in the Aleutian Islands relative to the 2000 census. However, it would be premature to conclude that the Gulf of Alaska population is recovering based on counts from a single year. The predominant cause of decline may have changed between the 1980s and 1990s. It is possible that minor factors during the 1980s have a larger relative impact now because the remaining population is much smaller. Observations made at one site may not apply to other areas or even to nearby rookeries. Research will be required at multiple sites to resolve whether survival is threatened by local, regional, or population-wide causes. **Finer-scale spatial analysis of Steller sea lion populations**

**and environmental conditions will be required to uncover potential region-specific determinants that are affecting sea lion survival.**

The more recent period of decline (1990-present) is the primary concern of this report because of the need to provide scientific advice for the design of management actions that do not jeopardize the continued survival of the western Steller sea lion population. Although limited in sample size, geographic range, and seasonality, recent measurements of sea lion condition and foraging activity indicate that the western stock is not nutritionally stressed and that individuals are not spending a disproportionate amount of time or energy in locating prey. Analysis of scat components provides evidence that dietary diversity is lower in the western range than in the eastern range, but this may represent opportunistic feeding patterns rather than a decrease in availability of preferred prey species. Additionally, the levels of groundfish biomass during the 1990s were large relative to the reduced numbers of sea lions, suggesting that there has been no overall decrease in prey available to sea lions, although it is still possible that localized depletion of some fish species may affect particular rookeries. **Existing data on the more recent period of decline (1990-present) with regard to the bottom-up and top-down hypotheses indicate that bottom-up hypotheses invoking nutritional stress are unlikely to represent the primary threat to recovery.**

Because the preponderance of evidence gathered during the current phase of the decline runs counter to expectations based on bottom-up hypotheses, the committee gave serious consideration to each of the top-down (direct mortality) hypotheses. All four hypotheses in the top-down category identify sources of mortality applicable to both the earlier and the current phases of the decline. What has changed since the 1980s is the potential impact of this mortality on the much smaller remaining population. Although killer whale predation may have had a significant impact on the historical population, continued predation, as well as illegal shooting, incidental takes by fishing gear, and subsistence harvests may have had a proportionately larger impact on the current depleted sea lion population. In the absence of other significant changes in the ecosystem, the intensity of bottom-up threats is expected to decrease as the sea lion population decreases, but top-down threats are often less dependent on population size. Sea lions remain easy targets for humans and marine predators because they congregate at rookeries and haulouts at certain times of the year. Similarly, sea lions may continue to get ensnared by fishing gear because of the ample banquet of food available around fishing operations. Attraction of killer whales to these same fishing vessels could increase the vulnerability of sea lions to predation. Identifying the most likely top-down hypothesis may depend on matching the different threats to the spatial patterns of sea lion population decline. Different hypotheses may apply to some but not all parts of the large geographic range of the western population. **Although no hypothesis can be excluded based on existing data, top-down sources of mortality appear to pose the greatest threat to the current population. Investigations of top-down sources of Steller sea lion mortality should be increased to evaluate the proportionate impact of these factors on the population decline.**

## MONITORING TO EVALUATE MANAGEMENT EFFICACY

**Although most evidence indicates that groundfish fisheries are not causing a range-wide depletion of food resources necessary to sustain the current western population of sea lions, there is insufficient evidence to fully exclude fisheries as a contributing factor to the continuing decline.** In some areas, fisheries may compete with sea lions for localized fish stocks, increase incidental mortality due to gear entanglement and associated injuries, disturb animals on haulouts, increase exposure to natural predators through attraction to fish catches, and provide motivation for continued illegal shooting of animals to mitigate lost catches and damaged fishing gear. Moreover, fisheries are one of the few human influences on the Steller sea lion's environment and hence are subject to regulation under the ESA. Therefore, restriction of fishing operations in sea lion habitat remains a reasonable response to the continuing decline of the endangered western population.

The committee has identified five general management options that might be taken to address the potential impacts of groundfish fisheries on sea lions and recommends monitoring priorities to assess the efficacy of each option. These options are evaluated with regard to their scientific potential for discerning the role of groundfish fisheries in the Steller sea lion decline. Each of these options would require continuation of the existing monitoring program (i.e., continued census of trend sites and collection of demographic data based on pup branding and resighting). The committee made the assumption that it is possible to craft each option so as to satisfy the requirements of the ESA. The five options are presented below.

1. *Wait and see, maintaining current closures indefinitely.* Recent management actions, including area closures, may be sufficient to reverse or reduce the rate of population decline. **Under this option the most valuable monitoring information would be derived from annual reference rookery and haulout counts and new demographic data from branded pups.**

2. *Eliminate direct fishery impacts with greatly expanded closures.* This

would require closing the Atka mackerel fishery in the Aleutians and reducing the main pollock fishing areas in the southern half of the eastern Bering Sea. **Under this option, monitoring of fish population dynamics, both locally and at the stock level, would be required to determine the effects of the fisheries on stock distribution and fish community composition.**

3. *Establish spatial management units consisting of two sets of closed and open areas where each treatment area is centered on a rookery.* The western population would be divided into management regions with at least two closed and two open rookeries per region. Because most monitoring activities are conducted at rookeries (pup counts, measurement of vital rates, juvenile tagging, etc.), it makes the most sense to use rookeries (rather than rookeries and haulouts) as the experimental units. Also, sea lions are thought to be more vulnerable near rookeries because of the age composition (presence of pups and juveniles) and because females must forage near the rookeries so that they can easily return to nurse their pups. The closed treatment units would be subject to fishery closures, and the open units would have sea lion-related fishery restrictions removed. **Under this option, the most critical monitoring needs would be detailed local Steller sea lion censuses and spatial analyses of fish population changes for each experimental unit in the overall design.**

4. *Implement a "titration experiment" where restrictions on fisheries (such as area closures) are increased progressively over time until a positive response is achieved.* This option is a variation on the strategy used during the 1990s. Fishery regulations continue to become more restrictive as long as the sea lion population continues to decline. **This approach requires monitoring of sea lion population trends, but results could be confounded by the lack of baseline data and natural environmental variability.**

5. *Micromonitor and manage localized interactions between sea lions and fisheries to reduce mortality where and when it occurs in the future.* **This option would require expansion of all basic monitoring activities (abundance, prey fields, mortality agent distribution) around key rookeries to pinpoint times and places of increased mortality so that appropriate management measures could be taken.** The expense of this program would be high because of a requirement for year-round continuous monitoring to allow detection of mortality events in all seasons and locations.

**To resolve questions about the impact of the fisheries on Steller sea lion survival, the preferred option is #3 because it is the only approach that directly tests the role of fishing in the decline.** Option #3 provides the benefits of an adaptive management experiment, reducing the possibility that regulation of the fishing industry is perpetuated without demonstrable benefit to the Steller sea lion population. Not only does the removal of all sea lion-related fishing restrictions in open areas create opportunities for the industry, it provides a contrasting management treatment necessary for a valid experimental comparison with closed areas. A careful evaluation of past fishing effort in the proposed experimental areas will be required to assess the amount of displaced fishing effort. Placement of open areas where fishing effort has historically been high would decrease the potential for negative impacts arising from shifting effort from the closed to the open areas.

Option #3 provides the setting necessary to carry out research studies on Steller sea lion behavior and performance in contrasting environments while controlling for common effects such as large-scale change in oceanographic regimes. This approach acknowledges that there is no best or precautionary policy because the origin of the decline is unknown. Hence, every segment of the population has an uncertain future with or without new restrictions on the fisheries. Multiple sites in various locations must be included in the experiment to control for site-specific variations in threats to the population. If there are multiple causal factors, such as food, predation, or fishing-related mortality, replication is critical to guard against incorrectly applying the results from any single treatment/control comparison to areas where the results would not apply.

Experimental treatment is a policy option that improves management and increases understanding of the interactions between fisheries and sea lions. Open areas restore opportunities for fisheries by removing restrictions; closed areas remove any potentially negative local impacts of fisheries on sea lions.

Although the incremental approach may be easier to implement, it contains two serious shortcomings. First, it cannot account for ecosystem change due to factors such as oceanographic regime changes. Hence, the efficacy of new management restrictions would not be distinguishable from environmental change that occurs on decadal timescales, confounding either positive or negative outcomes. Second, a false positive outcome would commit managers to prolong additional fishery restrictions without realizing significant improvement in the survival of Steller sea lions.

Listed below are several guidelines for implementing the spatial management units described under option #3:

- *Fished area (under normal management plans).* Design closures to minimize the displacement of fisheries to more distant, and less safe areas. The groundfish fisheries have been the focus of restrictions to protect sea lions based in part on the large amount of biomass removed, but the potential effects of other fisheries have not been as thoroughly examined. Hence, there are two basic experimental treatment options for area closures: (1) closure to groundfish fisheries only or (2) closure to all fishing. A positive response to the first treatment would measure the impact of the groundfish fisheries separately from the effects of other fisheries. A positive response to the second treatment would implicate fishing activities, but there would be uncertainty as to whether the response was due to exclusion of the groundfish fisheries or exclusion of another fishery—for example, herring or salmon. Closure of these areas to all fishing activity would provide the greatest contrast with the open areas for assessment of fishery-related effects on Steller sea lions. If only the groundfish fisheries are excluded from the closed areas, logbook data and as much observer coverage as possible should be obtained for other fisheries. Strict enforcement would be essential for correct interpretation of the effects of the closures.

- *Size and number of treatment areas.* The size of the closed areas depends on both fish movements and sea lion movements. The radius of the closure might range from 20 to 50 nautical miles (centered on a rookery). Replicates of each open/closed area comparison site will be required to assess the effects of environmental variability.

- *Timescale.* Some data gaps can be filled in less than 5 years (evidence of disease, localized fish depletion, improved estimates of direct mortality sources), but long-term monitoring (5 to 10 years) will be required to assess recruitment and mortality rates. If substantial numbers of Steller sea lions are taken as bycatch, open areas should be closed or fishing gears modified to prevent further decline of the population. This should apply to all fisheries that take sea lions as bycatch.

## RECOMMENDATIONS FOR RESEARCH AND MONITORING

**Research and monitoring should be directed toward measuring the vital rates and response variables most indicative of the status of the Steller sea lion population. This should include**

- *Population trends.* The current program for monitoring the juvenile and adult populations by aerial survey should be continued along with direct pup counts at selected rookeries.

- *Vital rates.* Vital rates have not been measured since the mid-1980s and urgently require updating. This should include measurements of fecundity, age at first reproduction, age distribution, juvenile survival, adult survival, and growth rates. Cooperative programs with subsistence hunters could provide reproductive data without additional mortality. Other parameters may be measured through increased effort in branding and resighting programs, requiring a commitment of resources for a period of time equivalent to the lifespan of the Steller sea lion.

- *Critical habitat.* Although the rookeries and haulouts of sea lions have been cataloged and described, the at-sea distribution of sea lions and related foraging activity are less well documented. Mostly this reflects the difficulty of collecting such data. The most valuable information comes from telemetry data, but analysis is constrained by the relatively small number of animals tagged, biases inherent in the recovery of data, and inaccuracies from inferring foraging activity based on swimming and diving behavior. Stomach telemetry tags that monitor temperature shifts associated with ingestion of prey should improve correlations of at-sea distribution with feeding. In conjunction with the analysis of Steller sea lions' at-sea activities, the activity and impacts of fisheries should be documented. Studies should be undertaken to determine if fisheries cause localized depletion of the various groundfish stocks through monitoring of fish distribution and density during the course of the fishing season with consideration of the need to distinguish these effects from natural changes in abundance. Designation of critical habitat should be revisited based on the results of the research proposed above.

- *Environmental monitoring.* Assessment of various ecological features of the sea lion environment will provide a broader context for evaluating sea lion population trends. These should include assessments of oceanographic conditions, plankton composition, forage fish abundance and distribution, seasonal migrations by groundfish, cephalopod abundance and distribution, and arrowtooth flounder interactions with groundfish (competition and predation). Also, monitoring for harmful algal bloom frequency and distribution through sampling of coastal waters will be valuable for assessing sudden mortality events. Biological sampling of sea lions should include testing for known marine mammal disease agents.

- *Predator feeding habits and population size*. Much more information is necessary to evaluate the impact of predation by killer whales and sharks on the continuing decline of the western population. Current evidence suggests that sharks are unlikely to be a major source of mortality based on distribution, limited diet data, and the relatively infrequent observations of shark wounds on sea lions. Better estimates of killer whale diet, population size, and distribution (including patterns of movement and habitat use) throughout Alaska are required to estimate potential predation mortality. In addition, observer programs should be instituted to record killer whale feeding behavior that may be different in different regions. Salmon shark and sleeper shark bycatch data from longline fisheries should be collected to assess shark abundance, and shark stomach contents should be examined to determine whether sea lions are a significant component of sharks' diets.

Most studies of Steller sea lions have been conducted in the summer, when sea conditions are favorable and it is relatively easy to work with females and pups on rookeries. However, this introduces a strong bias into the results because this season may not be the time when Steller sea lions are subject to increased mortality. The fate of juveniles remains a potentially pivotal question justifying the recent emphasis on their capture and tagging. In addition to increasing efforts directed toward year-round research at more accessible sites, remote observation methods such as satellite telemetry and video monitoring at rookeries and haulouts will be necessary to assess seasonal activity patterns. Although some research programs will yield data in a relatively short time (1 to 5 years), many of the variables most critical to assessing the efficacy of the various management regimes will take a minimum of 5 to 10 years before conclusive results are available. This is a consequence of the biology of sea lions; their long generation time means a slow population response and increased time required for assessing vital rates. Hence, it is even more urgent to develop and implement a prioritized cohesive research plan to address these information needs. Under an adaptive management scheme, the requirement to reduce jeopardy can be effectively coupled with a rigorous research program to reduce uncertainty about the causes of the ongoing decline of the Steller sea lion population.

# 1

# Introduction

Nothing endures but change.
—*Heraclitus.*

## CHALLENGES IN UNDERSTANDING THE CAUSE OF POPULATION DECLINES

The Gulf of Alaska, Aleutian Islands, and Bering Sea encompass a vast and spatially heterogeneous territory. The biological richness of this region has been exploited by humans for at least 5,000 years. Groundfish, such as Pacific halibut (*Hippoglossus stenolepis*)[1] and Pacific cod (*Gadus macrocephalus),* were first harvested in nearshore waters by Alaskan natives for subsistence. Beginning in the mid- to late 19th century, domestic fixed-gear (hook-and-line) fisheries began for Pacific cod, Pacific halibut, and sablefish (*Anoplopoma fimbria*). In the 1930s and 1940s, Japanese trawl fisheries developed for walleye pollock (*Theragra chalcogramma*) and flatfish (primarily yellowfin sole, *Pleuronectes aspera*) in the eastern Bering Sea. During World War II foreign fishing ceased. After the war, large multinational fisheries developed off Alaska. These included drift gillnet fleets for Pacific salmon; tangle net fisheries for crabs; longline fisheries for Pacific halibut, Pacific cod, sablefish, and Greenland turbot (*Reinhardtius hippoglossoides*) in the Bering Sea; and trawl fisheries for groundfish,

[1]Scientific names for species mentioned in the text are given in Appendix H.

herring (*Clupea pallasi*), and shrimp (*Pandulus* spp.). Foreign fleets largely comprised of small- to medium-sized (90 to 180 foot) trawlers and Danish seiners, some large factory stern trawlers (up to 270 feet), motherships for at-sea processing, and many support vessels were deployed to exploit groundfish resources off Alaska. Fishing effort shifted among species partly as a result of overfishing and changes in product demand by countries participating in the international groundfish fleet. Conflicts, including those among foreign trawlers and American fixed-gear vessels, resulted in increasing restrictions on foreign fleets. The Magnuson Fishery Conservation and Management Act of 1976 promoted domestic fishing by limiting the total allowable level of foreign fishing to be that portion of the optimum yield that was not expected to be harvested by domestic vessels. During the 1980s, foreign operations were converted to joint ventures between domestic catcher vessels and foreign processing vessels. In the 1990s the Alaska groundfish fishery became fully "Americanized."

The current domestic groundfish fisheries target walleye pollock (67% of total groundfish catch in 2000), Pacific cod, Atka mackerel (*Pleurogrammus monopterygius*), sablefish, and a variety of rockfish and flatfish species. In 2000 the groundfish fleet was comprised of 1,261 catcher vessels, 90 catcher processors, and 69 inshore processors and motherships. The majority of catcher vessels are less than 60 feet in length, but vessels in this size class accounted for only 20% of the exvessel value of groundfish harvested in 2000. The 15 catcher processor vessels that landed pollock in the eastern Bering Sea ranged in length from 201 to 376 feet.

Species composition and abundance in the North Pacific have undergone substantial variation over time. Small-mesh trawl survey data show a major shift in the relative abundance of benthic species from shrimps in the 1970s to groundfish in the 1980s (Anderson and Piatt, 1999). Increases in many flatfish and gadid species were particularly conspicuous (North Pacific Fishery Management Council, 2001a, 2001b) and overlapped with major declines of some crab populations, such as red king crabs (*Paralithodes camtschaticus*) throughout the Gulf of Alaska and Bristol Bay in the eastern Bering Sea (Zheng and Kruse, 2000). Variation also exists in the distribution and abundance of mammals and marine birds (Springer et al., 1999). Coincident with these changes, the western stock of Steller sea lions (*Eumetopias jubatus*) has undergone a persistent decline. The largest loss of animals occurred in the late 1970s and 1980s, but a more gradual decrease (about 5% per year) continued through the 1990s to 2000, with some signs of recovery in the Gulf of Alaska region seen in the recently released 2002 counts (see Figures 1.1 and 1.2). This decline, from a population estimated to be in the hundreds of thousands in the 1960s to roughly 30,000 in 2001, caused the National Marine Fisheries Service (NMFS) to list Steller sea lions as threatened in 1990 under the Endan-

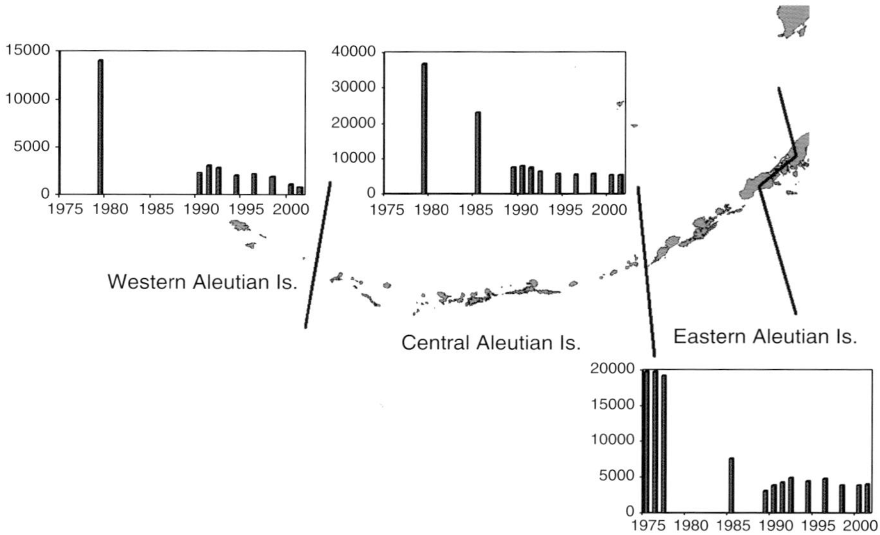


FIGURE 1.1 Steller sea lion population trends from 1975 to 2002 in the Aleutian Islands. (Maps and data from T.R. Loughlin, National Marine Fisheries Service, Seattle; data for 2002 from J.L. Sease, *Memo on Steller Sea Lion Survey Results, June and July 2002*, dated September 20, 2002.)

gered Species Act (ESA). In the face of continued decline in the west, this stock was proposed as endangered in October 1995 and listed as such in May 1997. It is not possible to determine whether the 1960s sea lion counts assessed the population above, at, or below the long-term average because reliable population data are not available to establish this type of baseline. A 19th-century description of Steller sea lions suggests that the population previously experienced large fluctuations in abundance, although again the cause is unknown but could be partially the result of human disturbance and hunting (see Appendix D).

Concerns about conflicts between Steller sea lions and commercial fisheries have a long history of scientific studies that started well before the listing of the western stock as endangered. Alverson (1992) provided a list of possible factors that contributed to the decline of Steller sea lions based on data from Loughlin (1987), and this list was revisited in the

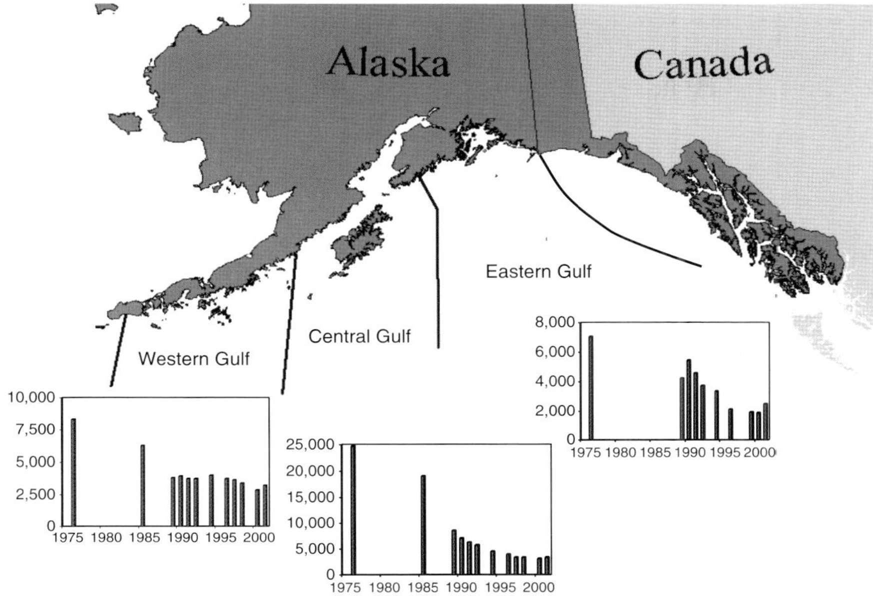


FIGURE 1.2 Steller sea lion population trends from 1975 to 2002 in the Gulf of Alaska. (Maps and data from T.R. Loughlin, National Marine Fisheries Service, Seattle; data for 2002 from J.L. Sease, *Memo on Steller Sea Lion Survey Results, June and July 2002*, dated September 20, 2002.)

Bowen et al. (2001) critique of NMFS's November 2000 Biological Opinion (BiOp #3). The hypothesized factors include changes in the species composition and abundance of Steller sea lion prey, disease, toxins, killer whale predation, intentional and incidental (or illegal) killing of Steller sea lions, and regime shifts in the physical environment. From the first documentation of a population decline in the 1980s to fiscal year 2001, there has been an escalation of funding for research aimed at identifying plausible causes of the decline (see Appendix E). In fiscal year 1992, research funding was a modest $1.43 million, compared with $43.15 million in fiscal year 2001 and $40.145 million in 2002 (DeMaster and Fritz, 2001; Ferrero and Fritz, 2002).

Every ecosystem presents a multitude of complex processes and interactions that defy straightforward description and explication. With respect to this particular ecosystem, spatial structures are important features, whether at the scale of sea lion rookeries and haulouts, oceanic fronts, or entire oceanic basins. Spatial ecology (Tilman and Kareiva, 1997) provides a conceptual perspective that helps to unite existing data on rookery-specific analyses and pollock stock dynamics and movement (Shima et al., 2002). The challenge here is to bring together these spatial perspectives—regional and local—in a structured approach so that it is possible to generate and test hypotheses about the causes of the continuing sea lion population decline.

At present there is insufficient data to conclusively identify the cause or causes of the decline. However, the available information can be organized both to identify the most fruitful avenues for future research and to evaluate the likely efficacy of current and future management actions. This weight of evidence approach, described in Box 1.1, provides an analytical framework for timely decision making when scientific certainty is constrained by limited historical data, high system variability, and extremely challenging field conditions.

## THE POLICY CONTEXT

Is it possible to reach definitive answers about the cause of the Steller sea lion decline? Complex systems, by definition, are not amenable to simple description and analysis. Not only is the Steller sea lion situation complex, there is a dearth of information about both the ecosystem and the sea lions before and during the decline that could be used to reduce uncertainty about the cause or causes of the population decrease. When that complexity includes an important economic activity—in this case the groundfish fisheries—policy made in the absence of certainty often results in various legal challenges.

Science proceeds by seeking to rule out hypotheses that do not seem credible and by identifying those hypotheses that appear plausible and supportable. In this process there is often a reluctance to promote specific hypotheses as causal, but there is often less reluctance to rule out alternative hypotheses as unimportant. Hence, one focus of this report is to evaluate evidence both for and against the various hypothesized causes of the Steller sea lion decline.

Several pieces of legislation form the context for the Steller sea lion controversy. The Magnuson-Stevens Fishery Conservation and Management Act of 1996 calls for the protection of marine ecosystems, assuring that "irreversible or long-term adverse effects on fishery resources and the marine environment are avoided"(16 U.S.C. 1802 (5)). The Marine Mammal Protection Act states: "The primary objective of (marine mammal) management should be to maintain the health and stability of the marine ecosystem" (16 U.S.C. 1361). The National Environmental Policy

**BOX 1.1**
**Nature of Evidence and Standards of Inference**

The strength of scientific inference varies greatly depending on both the nature of the problem and the availability of relevant reliable data. Strong inferences can be made when there is a hypothesis testable by experimentation. The standard procedure is to establish a decision criterion (usually termed the critical value) against which the hypothesis is either accepted or rejected based on the experimental result. Usually the decision criterion is chosen to reduce the probability of incorrectly accepting the hypothesis. This procedure is the preferred approach of scientific inquiry.

In natural resource management and conservation, there may be obstacles to experimentation. Achieving this high standard of inferential rigor might require more time or resources than would be practical for management of the affected resources. In some cases, the necessary experiments may even be deemed unethical or illegal or the complexity of the system may reduce the likelihood of obtaining unequivocal results. In these situations, information from various sources can be assembled to make a weight of evidence argument among the competing hypotheses. In general, the strength of a weight of evidence inference depends on the number of independent lines of evidence and the relevance of each. The strength of the inference is typically weaker when there are a number of plausible alternative hypotheses and the problem may be caused by multiple factors.

No fewer than eight hypotheses have been proposed to explain the Steller sea lion decline (see Chapter 6). A wide array of data is available on sea lions and their associated ecosystems, including spatiotemporal patterns of population change, information on body condition and foraging behavior, fish stock assessments and fishing effort, population trends in other co-occurring species, and a variety of historical records. Given various inferences about how these data relate to each hypothesis, a weight of evidence assessment can be established. Furthermore, hypotheses that are inconsistent with multiple lines of evidence can be ranked as less likely than hypotheses consistent with existing evidence. As with all types of analyses, the robustness of a conclusion is based on the quality of the data and potential biases introduced during the analysis.

Act seeks "efforts which will prevent or eliminate damage to the environment and biosphere" by requiring federal agencies to "identify and develop methods and procedures that will insure that presently undocumented environmental amenities and values may be given appropriate consideration in decisionmaking"(42 U.S.C. 4321). The ESA provides "a means whereby the ecosystems upon which endangered species and threatened species depend may be conserved [and] a program for the conservation of such endangered species and threatened species"(16 U.S.C. 1531 (b)). While these legislative mandates approach protection from different perspectives, careful management of human use of living marine resources remains their overarching goal.

Under the ESA, federal agencies are required to ensure that their actions, or actions authorized or funded by them, are not likely to jeopardize the survival or recovery of protected species or damage their critical habitat. Section 7 of the ESA requires that when an action may affect a listed species or its critical habitat, the federal agency conducting or authorizing that action (the action agency) must consult with the federal agency charged with overseeing recovery efforts for the listed species (the expert agency). In cases where a federally managed fishery may interfere with the survival or recovery of certain marine mammals (seals, sea lions, porpoises, and whales), NMFS is both the action and the expert agency. These responsibilities are segregated within NMFS between the Office of Sustainable Fisheries, which has responsibility for reviewing the fishery management plans, and the Office of Protected Resources, which has responsibility for implementing ESA regulations for listed species under its jurisdiction.

The consultation process allows for the possibility that the agency will be unable to develop reasonable and prudent alternatives (RPAs) that will remove jeopardy (50 *CFR* 402.15[h], 2001). The biological opinion requires consideration of the extent to which listed species are vulnerable to "incidental takes" as an unintended consequence of a particular action under consideration. These statements on incidental takes must include reasonable and prudent measures to minimize the probable impacts arising from such takes and must set forth terms and conditions that will minimize probable impacts. On the basis of this information, the federal action agency (NMFS Office of Sustainable Fisheries) will determine whether or not to proceed with the proposed action (in this case the groundfish fishery in the Bering Sea/Aleutian Islands and the Gulf of Alaska (Figure 1.3).

While neither "jeopardy" nor "adverse modification" is defined in the ESA, the Code of Federal Regulations defines jeopardy as follows: "Jeopardize the continued existence of means to engage in an action that reasonably would be expected, directly or indirectly, to reduce appreciably the likelihood of both the survival and recovery of a listed species in the wild by reducing the reproduction, numbers, or distribution of that species" 50 *CFR* 402.02, p. 38 (2001).

Scientists and managers cannot ensure specific outcomes of regulatory actions even when legislation requires a particular outcome, such as the preservation of a particular species. In some cases, federal agencies have been forced to relinquish discretion in the matter of endangered species policy to the courts. If the public is not satisfied as to whether or not an agency's actions seem reasonable in light of its obligations to an endangered species, the determination is made by the courts. This reflects the role of courts in the U.S. constitutional system of judicial oversight of

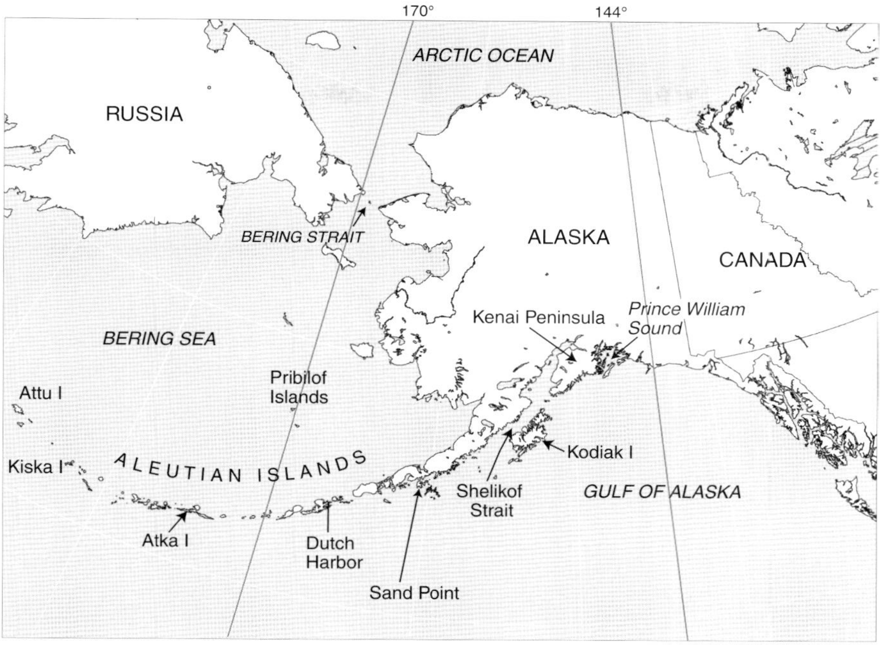


FIGURE 1.3 Map of Alaska and surrounding waters.

legislative intent and language and executive branch action in response to that language.

If a jeopardy decision is rendered, the agency must issue a set of RPAs that would remove the finding of jeopardy. These RPAs must meet four conditions: (1) they must be identified during formal consultation and must be implemented in a manner consistent with the intended purpose of the action; (2) they must be implemented in a manner consistent with the agency's legal authority and jurisdiction; (3) they must be economically and technically feasible; and (4) they must be actions that the agency believes will ameliorate the original finding of jeopardy to the listed species or its habitat. A successful RPA must meet all four of these conditions. When uncertainty exists, the benefit of the doubt must be given to the listed species (*Greenpeace v. National Marine Fisheries Service,* 55 F. Supp. 2d, at 1262).

The third condition for a successful RPA is that actions taken should be economically and technically feasible. This does not mean that an agency is required to balance benefits to the endangered species against costs falling on an industry. To do so would be "fundamentally inconsistent with the purposes of the ESA and with the case law interpreting the ESA" (55 F. Supp. 2d, at 1267). Instead, case law suggests the application of lexicographic choice. In lexicographic choice one first satisfies the paramount goal and then chooses among alternatives consistent with that goal. In the current context, lexicographic choice requires that the NMFS develop sets of reasonable and prudent alternatives that comply with the ESA mandate of no jeopardy. An evaluation of benefits and costs can then be conducted to select a preferred option from among these RPAs.

## HISTORY OF ESA LISTINGS AND COURT CHALLENGES

As part of the authorization of the fishery management plans for the commercial groundfish fisheries in the Bering Sea/Aleutian Islands and the Gulf of Alaska regions, NMFS summarized the consultation in a biological opinion as required under Section 7 of the ESA. The purpose of the biological opinion is to ascertain if the groundfish fisheries, as prosecuted under the fishery management plans, are likely to imperil the continued existence of Steller sea lions (and other listed species) or are likely to destroy or adversely modify critical habitat. In April 1998, Greenpeace filed a complaint in U.S. District Court that NMFS had failed to revise the environmental impact statement relating to federal groundfish fisheries in Alaska and had violated the ESA because the biological opinions regarding the impacts of these fisheries on sea lions were inadequate. In the biological opinion issued in December 1998 (known as BiOp #1), NMFS concluded that the groundfish fisheries, excepting pollock, were unlikely to cause harm to listed species. In the case of the pollock fishery, there was a finding of jeopardy based on competition between the fishery and sea lions for pollock. In response to this finding, a set of RPAs was developed in consultation with the North Pacific Fishery Management Council (NPFMC) that spread fishing effort out spatially and temporally and closed the Aleutian Islands management area to pollock fishing. These restrictions were implemented in the 1999-2000 fishery management plans. After the RPAs went into effect in January 1999, NMFS issued another biological opinion (BiOp #2), which analyzed the effects of the entire groundfish fishery management plan on sea lions and found no jeopardy from the pollock fishery based on a review of the total allowable catch levels proposed for the Gulf of Alaska and Bering Sea/Aleutian Islands management areas.

Greenpeace filed suit in response to the new biological opinion and on July 9, 1999, U.S. District Court Judge Thomas Zilly found the RPAs to be arbitrary and capricious because there was no explanation of how the proposed restrictions mitigated jeopardy for the pollock fishery. He also found that the environmental impact statements were inadequate and directed NMFS to prepare a more comprehensive analysis of the Bering Sea, Aleutian Islands, and Gulf of Alaska groundfish fisheries (*Greenpeace v. National Marine Fisheries Service*, 80 F. Supp. 2d 1137 WD. Wash., 2000). In January 2000, Judge Zilly ruled that the "no jeopardy" finding in BiOp #2 was inadequate under ESA because it only considered the total allowable catch levels for individual groundfish fisheries and failed to consider the combined and cumulative impacts of all groundfish fisheries on sea lion populations. Based on the January ruling, Greenpeace filed for an injunction prohibiting groundfish trawling in sea lion critical habitat until a new comprehensive biological opinion was prepared by NMFS. The injunction was granted in July and implemented in August 2000.

NMFS released the revised biological opinion on November 30, 2000 (BiOp #3). It concluded that Steller sea lion populations are jeopardized by the Alaska groundfish fisheries, including Atka mackerel, Pacific cod, and pollock, due to competition for prey and modification of prey distribution in critical habitat. This revised biological opinion found jeopardy with regard to pollock even under the restrictions imposed by the 1999 RPAs. The opinion included a comprehensive set of new RPAs that incorporated adaptive management to assess the efficacy of the groundfish restrictions. The western population was divided into 13 management areas designated as either open—with fishing allowed under the 1999 restrictions—or closed—with no fishing allowed in critical habitat. However, these new regulations were resisted as being too costly to the groundfish fisheries, and Alaska Senator Ted Stevens attached an amendment to the December 2000 omnibus appropriations bill that delayed full implementation of the RPAs and provided the NPFMC with an opportunity to develop an alternative set of RPAs. In addition, the amendment provided $30 million for economic relief to offset losses incurred by sea lion protection measures, $28 million for research on the causes of the decline of sea lions, and $2 million for scientific review of BiOp #3, including the review by the National Academy of Sciences that is the subject of this report.

In February 2001 the NPFMC appointed an RPA committee to develop alternatives to the RPA in BiOp #3 that removed the potential for jeopardy from the pollock, Atka mackerel, and Pacific cod fisheries but had less impact on the fishing industry and associated communities. In June the RPA committee proposed an alternate set of measures that discarded the earlier adaptive management approach and used new telemetry data to justify restricting fishing primarily in the first 10 nm of the 20-nm radius delineating critical habitat areas. The telemetry data suggest that sea lions spend most of their time at sea within 10 nm of the rookeries. The revised RPA assumes that the telemetry data reflect the foraging behavior of sea lions, and hence 10 nm delimits the zone with the maximum likelihood of competition with the fisheries. By moving most fishing activities beyond 10 nm, with some further restrictions between 10 and 20 nm, the RPA committee was able to reach the same theoretical reduction of jeopardy as in BiOp #3. In August 2001, NMFS released BiOp #4, which evaluates the new RPA measures and includes a supplemental environmental impact statement that compares the various RPA measures. NMFS concludes in BiOp #4 that the June 2001 RPAs provide adequate protection for Steller sea lions with regard to the groundfish fisheries.

## SCOPE AND ORGANIZATION OF THE REPORT

In response to the congressional request for a review of the Steller sea lion decline and the Alaska groundfish fishery, the Ocean Studies Board and the Polar Research Board of the National Academies agreed to undertake a study of the issue, which was funded through the NPFMC. A committee of experts in marine mammal biology, marine ecology, and fisheries science was convened in the summer of 2001 to address the issues of concern listed in the study's statement of task (see Box 1.2). See Appendix A for committee biographies.

The committee held three public meetings, two in Seattle and one in Anchorage, to receive input from the NPFMC, NMFS, academic scientists, fishermen, Alaska natives, environmentalists, and other concerned members of the public.

The report is organized to provide a rationale for analyzing the causes of the sea lion decline based on what is known about the demographics of

**BOX 1.2**
**Statement of Task**

This study will examine interactions between Alaska groundfish fisheries and Steller sea lions (*Eumetopias jubatus*) and the role of these fisheries in the evolving status of the sea lion population. The focus of the study will be (1) the status of current knowledge about the decline of the Steller sea lion population in the Bering Sea and Gulf of Alaska ecosystems; (2) the relative importance of food competition and other possible causes of population decline and impediments to recovery; (3) the critical information gaps in understanding the interactions between Steller sea lions and Alaska fisheries; (4) the type of research programs needed to identify and assess potential human and natural causes of sea lion decline; and (5) the components of an effective monitoring program, with yardsticks for evaluating the efficacy of various management approaches.

the population and the known linkages of Steller sea lions, human activities, and various features of food webs in the North Pacific region. Chapter 2 describes the general features of the environmental setting based on the climatic, oceanographic, and biological features of the region defined by the range of the western Steller sea lion population. An accounting of sea lion mortality based on population and ecosystem models is presented in Chapter 3 to evaluate how much of the sea lion decline cannot be explained by known factors. This analysis also helps identify the types of mortality that could most readily explain the observed pattern of sea lion population decline. Readers unfamiliar with Steller sea lion biology may prefer to consult the review of sea lion biology presented in Chapter 4 prior to reading the modeling discussion in Chapter 3. Chapter 5 describes the North Pacific commercial fisheries in reference to potential interactions with Steller sea lions. These summaries provide the basis for the weight of evidence approach used in Chapter 6 to evaluate each of the eight major hypotheses proposed to explain the Steller sea lion decline. Identification of critical information gaps, research approaches, and recommendations for future monitoring programs are presented in Chapter 7.

# 2

# The Environmental Setting

## THE PHYSICAL ENVIRONMENT

To place the decreased abundance of Steller sea lions (*Eumetopias jubatus*) in context with changes in the sea lion's environment, this chapter describes salient features of the physical and biological regimes of the North Pacific. Steller sea lions in the endangered western stock live throughout the Gulf of Alaska (west of ~144° W), Aleutian Islands, and eastern Bering Sea (Figure 2.1). Climate variability and oceanic responses to climate affect biota, including sea lions, through a number of pathways and mechanisms (Francis et al., 1998; Schumacher and Alexander, 1999; Schumacher et al., in press). Because of severe limitations in data and understanding, researchers face a number of challenges when attempting to identify cause and effect relationships between climatic-environmental changes and changes in marine species such as Steller sea lions (see Benson and Trites, 2002, for a comprehensive review). In general, changes in the physical environment may affect the abundance and types of prey available to Steller sea lions through nutrient-phytoplankton-zooplankton dynamics and responses of higher-trophic level biota such as forage fish and groundfish. Physical perturbations, particularly those in upper-ocean temperature, could affect Steller sea lions by changing the distribution and abundance of sea lion prey, competitors, or predators.

The physical environment in the western range of the Steller sea lions is profoundly influenced by atmospheric phenomena (e.g., wind stress, heat flux, precipitation). Frequent fall and winter storms (low-pressure

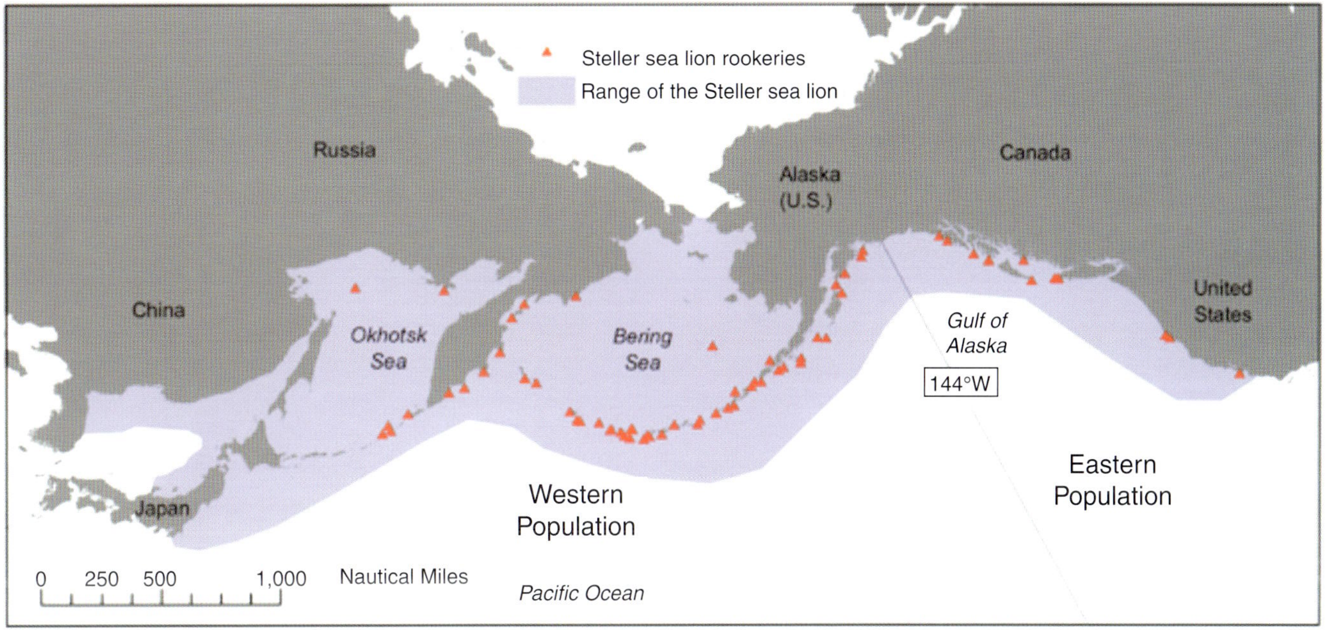


FIGURE 2.1 The range (in Purple) and rookeries (in red) of Steller sea lions. 144° W defines the boundary between the eastern and western populations.
SOURCE: National Marine Fisheries Service, Alaska Fisheries Science Center (*www.afsc.noaa.gov/stellers/range.htm*).

systems) passing across the Aleutians and midlatitudes of the North Pacific produce the climatological Aleutian Low pressure cell. Averaged over one to several months, the Aleutian Low pressure pattern exhibits strong intraseasonal, interannual, and interdecadal variability (Trenberth, 1990; Trenberth and Hurrell, 1994). The intensity of the fall/winter Aleutian Low is correlated with El Niño-Southern Oscillation variations (primarily on a 2- to 7-year timescale) and also varies as an element of the Pacific Decadal Oscillation (PDO) at timescales as long as 50 to 70 years (Minobe, 1997). The Aleutian Low varies in association with the Arctic Oscillation over a broad spectrum of timescales (interannual to interdecadal; Overland et al., 1999).

In the North Pacific and Bering Sea, 20th-century climatic regimes persisted for periods ranging from 10 to 30 years (Francis and Hare, 1994; Ware, 1995; Mantua et al., 1997; Minobe, 1997). Tree-ring reconstructions for coastal Alaska and North Pacific surface temperatures provide evidence of multidecadal climatic regime shifts over the past three centuries, though the amplitude of this variability was weak in the mid-1980s (e.g., Ingraham et al., 1998; D'Arrigo et al., 2001; Gedalof and Smith, 2001). Studies of tree-ring records (Ingraham et al., 1998) and nitrogen isotopes in salmon remains (Finney et al., 2000) have identified periodic changes in the climate and biota at least as far back as the 1500s. The PDO represents one aspect of a regime shift, which is manifested in the patterns of sea surface temperature throughout the Pacific Ocean north of 20° N. Recent regime shifts occurred in 1976-1977, 1989, and likely in the late 1990s (see Figure 2.2). Despite strong year-to-year climate variability in both the North Pacific and Bering Sea, there remains compelling evidence for coherent interdecadal variations in the large marine ecosystems of the North Pacific and Bering Sea (at least as indicated by collections of fishery and ecosystem survey records from 1965 to 1997). Hare and Mantua (2000) speculate that aspects of North Pacific and Bering Sea ecosystems have responded nonlinearly to "noisy" external forces like those associated with strong year-to-year climate variations that are ubiquitous in midlatitudes.

In subarctic regions, sea ice is an important feature that links atmospheric changes to the ocean. Air-sea heat exchange and wind stress dictate the formation, transport, and persistence of sea ice. In the Gulf of Alaska and the Aleutian Islands, sea ice forms in small areas: the greatest coverage occurs in Cook Inlet (Wilson and Overland, 1986) and plays a minor role in local ecosystem dynamics. In the eastern Bering Sea, variations in sea ice extent, time of advance, and subsequent disappearance either by melting in place or being blown away are striking physical phenomena (Schumacher and Stabeno, 1998). The strongest signal occurs at interannual periods, although changes also occur during regime shifts

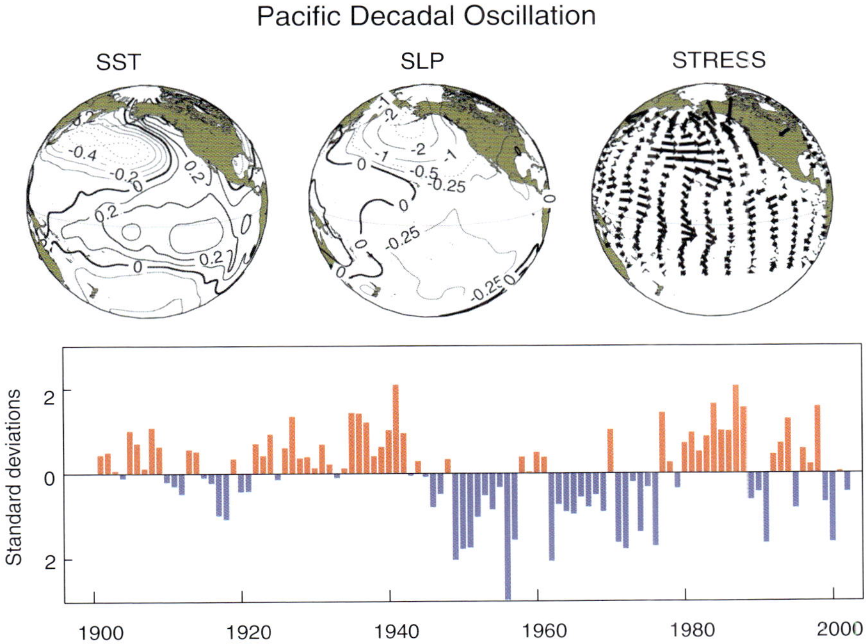


FIGURE 2.2 Anomalous climate conditions associated with warm phases of the Pacific Decadal Oscillation (PDO). Sea surface temperature (SST) values are given in degrees Celsius, sea level pressure (SLP) is in millibars, and surface wind stress direction and intensity are indicated by arrows where the longest wind vector represents a pseudostress of 10 $m^2/s^2$. The lower graph presents November-March average values of the PDO index during the 20th century. Actual anomaly values for a given year at a given location are obtained by multiplying the climate anomaly by the associated index value (Reprinted from Mantua and Hare, 2002, with kind permission of Kluwer Academic Publishers).

(Brodeur et al., 1999; Stabeno and Overland, 2001). The changes in sea ice characteristics have profound impacts on biota (Schumacher et al., in press). A bloom of phytoplankton is associated with the spring sea ice and accounts for 10 to 65% of the total annual primary production (Niebauer et al., 1995). In years when sea ice is either not present or retreats before adequate light is available (early March), the conventional spring bloom is delayed until late spring (Stabeno et al., 2001). The newly described oscillating control hypothesis (OCH) relates ecosystem changes to changes in atmospheric and oceanic features of the southeastern Bering Sea (Hunt et al., 2002). In essence, the OCH states that during cold physical regimes the ecosystem is driven in a bottom-up manner, whereas during warm regimes it is in a top-down mode. Under the OCH, cold regimes are characterized by late retreat of sea ice and low water column temperatures. At low temperature, survival of fish eggs is reduced and there is less zooplankton prey for larval fish, resulting in reduced recruitment to adult populations of pollock and other fishes.

Bottom-up processes generally dictate the flow of energy through the ecosystem during a cold regime. Because low water column temperatures can change the distribution of some forage fish species, the OCH explains why Steller sea lions, other pinnipeds, and piscivorous seabirds may thrive even under cold conditions if forage fish become more available as prey. Warm regimes occur during years when sea ice is either absent or retreats before there is sufficient sunlight to initiate an under-ice bloom. In this case, the spring bloom occurs later, water column temperatures are warmer, and higher zooplankton production provides prey for larval and juvenile fish. This should favor recruitment and strong year classes of pollock and other piscivorous fish.

There are years when the observed regime conditions (e.g., 1976, cold with low recruitment; 1996, warm with high recruitment) and estimated pollock production fit the outcomes predicted by the OCH; however, there are also years that do not fit this pattern (e.g., 1987, warm with low recruitment; 1992, cold with high recruitment). The following scenario explains how the OCH could affect the abundance of fish stocks through top-down control by predators. Adult pollock eat age 1 and younger fish and together with predation by other fish (e.g., arrowtooth flounder and Pacific cod) may depress the size of new year classes (Livingston and Methot, 1998; Livingston et al., 1999).

When there is a sequence of warm regime years, recruitment will be above average and the populations of adult predatory fish will increase to a point where the control of future year-class strength will be mainly a top-down process. In the Gulf of Alaska, for example, expansion of the arrowtooth flounder population correlates with a severe decline in the pollock population (Bailey, 2000). If predation on forage fish increases, zooplankton will become available for other populations (e.g., jellyfish, salmon, baleen whales). Farther up the food web, the reduced abundance of forage fish could affect the feeding success and survival of pinnipeds and piscivorous seabirds.

Although decadal timescale or regime shifts have been observed in the Bering Sea (e.g., Stabeno et al., 2001; Minobe, 2002), much of the variance is in the annual cycle (Royer et al., 2001). The OCH addresses the impact of annual changes in the following manner. If a cold year occurs during a warm regime, the existing stock of predatory adult fish will maintain top-down control of prey fish, reducing recruitment of the next year class. During a cold regime, stocks of piscivorous predators (e.g., pollock) will be relatively low; hence, recruitment will mostly depend on bottom-up processes. One or two warm years could yield strong year classes of piscivores that could then switch the system from bottom-up to top-down control by predation. Variations in forage fish and pollock year class strength could also occur within a regime if predators become redistributed.

In addition to the effects on bottom-up forcing of energy flow in the pelagic ecosystem, ice and its attendant cold pool of water directly influence distributions of higher-trophic-level biota such as marine mammals by providing habitat (Ohtani and Azumaya, 1995; Wyllie-Echevarria and Wooster, 1998; Brodeur et al., 1999).

The coastline of the Gulf of Alaska is marked by a nearly continuous mountain barrier that results in topographic uplift high winds and substantial precipitation (Wilson and Overland, 1986; Royer, 1998; Royer et al., 2001; Stabeno et al., in press). The continental shelf relevant to the western stock of Steller sea lions is generally about 150 km wide with many troughs and underwater canyons. Unlike the shallow eastern Bering Sea shelf, in the Gulf of Alaska depths greater than 150 m often occur close to the coast. Numerous islands provide habitat for the Steller sea lion. The Alaska Current flows westward and becomes a strong boundary current, the Alaskan Stream off Kodiak Island (Reed and Schumacher, 1986). The dominant current over the shelf is the Alaskan Coastal Current (ACC), which flows generally westward along the coastline (Schumacher and Reed 1986; Stabeno et al., 1995; Royer, 1998; Stabeno et al., in press) and is generated by a combination of wind-driven convergence along the coast and freshwater runoff. The ACC is a continuous feature westward to Unimak Pass, where it flows northward to the Bering Sea shelf (Reed and Schumacher, 1986). The transport of nutrients from the oceanic waters occurs in the Shelikof Strait with the mixing of fresh water and seawater as the ACC flows seaward (Reed et al., 1987). Other mechanisms that renew nutrients on this shelf include flow along the deep troughs, oceanic eddies, and an Ekman flux (Stabeno et al., in press). The latter process results from the mean winds generating upwelling in the central North Pacific Ocean that is then driven toward the Alaskan coastline in the upper or Ekman layer of the water column.

The Aleutian Islands form a porous boundary between the North Pacific Ocean and the Bering Sea, with a narrow continental shelf. Strong tidal currents mix the northward-flowing waters, which can then provide nutrients to the euphotic zone. Adjacent to the Aleutian Islands, the dominant current to the south is the Alaskan Stream, which provides relatively fresh surface and warm subsurface waters to the Bering Sea. The flow northward through Amchitka and Amukta Pass is primarily responsible for supplying the Aleutian North Slope Current (Reed and Stabeno, 1999), which flows eastward along the northern side of the Aleutian Islands and provides the main source of the Bering Slope Current. The importance of these currents to chemical and planktonic material is threefold. First, they provide transport from oceanic to shelf waters, a feature potentially important for pollock that spawn near Bogoslof Island. Second, their eddies can temporarily provide a habitat that favors survival of larvae (Schumacher and Stabeno, 1994). Third, their relatively warm temperature characteristics are potentially important to fish stocks by increasing egg maturation rates and thus reducing time for predation on that life history stage (Reed, 1995).

The continental shelf of the southeastern Bering Sea is broad (~500 km) and relatively shallow (<180 m) with canyons incising the shelf break. Compared to the Gulf of Alaska, the bathymetry of this shelf is featureless. Unlike the Gulf or Aleutian Islands, circulation over the shelf proper is generally sluggish. There is a moderate flow toward the northwest over the outer shelf in the vicinity of the 100-m isobath, and a weaker feature exists around the perimeter of the eastern shelf flowing east and then northward in the general vicinity of the 50-m isobath (Schumacher and Stabeno, 1998). Tidal currents mix the waters of the coastal domain and create a bottom mixed layer over the remainder of the shelf. Low-frequency currents around some of the islands and in submerged canyons are generated as water masses encounter submerged topographic features (Schumacher and Stabeno, 1998; Kowalik, 1999; Stabeno et al., 1999).

## THE BIOLOGICAL ENVIRONMENT

It is now generally recognized that ecosystems can shift between different states and that these shifts often involve gradual changes in a system until a threshold is reached and reorganization of the ecosystem occurs (Francis and Hare, 1994; Anderson and Piatt, 1999; Scheffer et al., 2001). Recent studies indicate that high-latitude Arctic ecosystems may be particularly vulnerable to these regime shifts, with these transitions potentially happening every few decades, especially when they are associated with decreases in sea ice extent (Aagaard and Carmack, 1989; Manabe et al., 1991; Lynch et al., 1995; Hu et al., 1999). The allocation of carbon fixation to respiration, storage, and export, as mediated by coupling of pelagic and benthic carbon cycles, is of major interest (Walsh et al., 1989; Walsh, 1995; Rivkin and Legendre, 2001). Although hemispheric climate variability has been tied to regime shifts, threshold responses in the regional Gulf of Alaska and/or Bering Sea biological ecosystem are more elusive yet critical to understanding the environmental factors potentially acting on the decline of Steller sea lion populations in the region.

Large-scale climate and ecosystem changes have been documented in the North Pacific and Gulf of Alaska (Francis et al., 1998; Hare and Mantua, 2000) and the Bering Sea (Mantua et al., 1997; Minobe, 1997; Scheffer et al., 2001) in 1925, 1947, 1977, and 1989 (Beamish et al., 1999; Overland et al., 1999). It has been suggested that these changes in biological and physical variables have ecosystem-wide consequences as seen in changes of food web structure and carbon cycling (Beamish et al., 1999; Hare and Mantua, 2000). These shifts have had widespread and important impacts on commercial fisheries, sea birds, marine mammals, and Alaska natives who harvest these resources (National Research Council, 1996; Callaway et al., 1998; Criddle et al., 1998).

### Changes in North Pacific and Gulf of Alaska Biology

Oceanic and atmospheric interactions in the North Pacific and Gulf of Alaska have a direct influence on water column productivity. Intensification of regional winds is believed to enhance primary productivity by increasing the divergence of surface water in the center of the Alaska Gyre, enhancing the upwelling of nutrient-rich water. Surface water warming and enhanced stability would keep the phytoplankton in more favorable light conditions in the euphotic zone, thus potentially boosting primary production when light is the limiting factor. Copepod biomass also increased after the 1976-1977 regime shift, compared to the 1950s to mid-1970s, likely due to an increase of the Alaska Gyre and an intensification of the Aleutian Low (Brodeur and Ware, 1992).

Historical records tracking aspects of Pacific marine ecosystems suggest a strong association between the PDO variability and Pacific salmon production (Beamish and Bouillon, 1993; Beamish et al., 1999; Hare et al., 1999), Pacific sea birds (Vandenbosch, 2000), Alaska groundfish and zooplankton production in the central and eastern North Pacific (Francis et al., 1998; Hollowed et al., 1998), and Gulf of Alaska marine species assemblages (Anderson and Piatt, 1999), to name just a few (see Figure 2.3). The small-mesh Gulf of Alaska trawl survey data analyzed by Anderson and Piatt show an apparent major redistribution in the relative abundance of species closely associated with the 1977 regime shift (see Figure 2.4).

As the relatively cool era of the early 1970s switched to a relatively warm era (in Gulf of Alaska sea surface temperatures and ocean bottom temperatures and wintertime coastal Alaska air temperatures), data on catch per unit effort indicate that shrimp and small pelagic fish became scarce concurrently with a dramatic increase in the biomass of gadids and flatfish (Ronholt et al., 1978; Alverson, 1992; Anderson and Piatt, 1999; Mueter and Norcross, 2000).

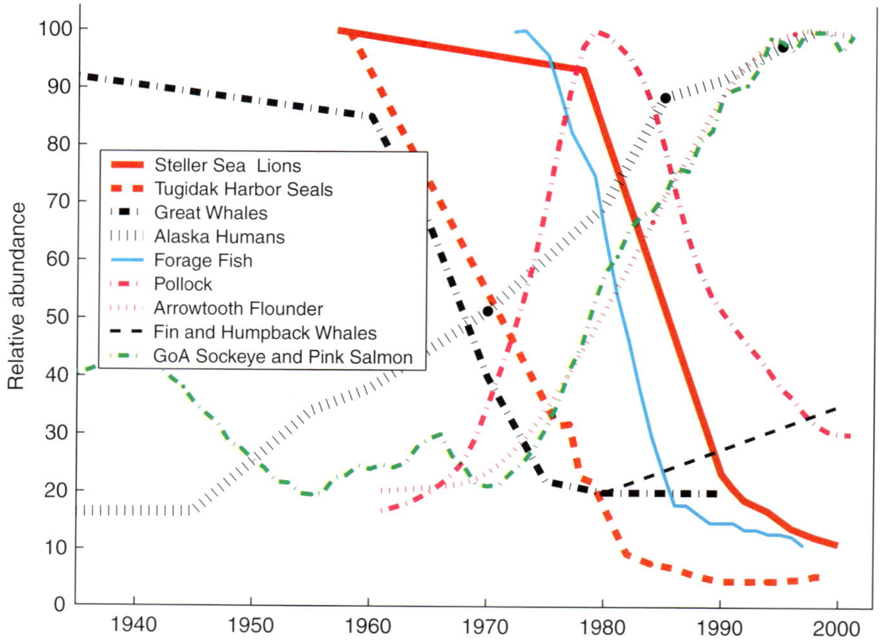


FIGURE 2.3 Gulf of Alaska ecosystem trends: 1945-2000. Schematic diagram indicates the relative abundance of select marine mammals, fish, and shellfish in the Gulf of Alaska. To clarify multiyear trends a 10-year running average was applied to groundfish, forage fish, and Tugidak harbor seal molt count data.
SOURCES: Data for blue, fin, sei, sperm, and humpback whales (key = Great Whales) came from National Research Council (1996). Fin and humpback whale population trends were updated with data from Robyn Angliss and Sue Moore from NMFS National Marine Mammal Lab, Seattle, WA, personal communication. Data for groundfish (pollock and arrowtooth flounder) came from NMFS 2001 stock assessments (*www.refm.noaa.gov/docs/).* Steller sea lion data came from NMFS stock assessments (*www.refm.noaa.gov/docs/*). Human population came from Alaska census data (*www.labor.state.ak.us/research/popover.htm*). The 1958 data point for Tugidak harbor seal beach molt counts came from Douglas DeMaster (National Marine Fisheries Service, Marine Mammal Lab, Seattle, personal communication, 2002). All other data for harbor seals came from L.A. Jemison and G.W. Pendleton, *Harbor Seal Population Trends and Factors Influencing Counts on Tugidak Island, Alaska*, an unpublished manuscript provided by Bob Small of the Alaska Department of Fish & Game, November, 2002. Data for Gulf of Alaska forage fish catch per unit effort was estimated from the NMFS's small-mesh trawl survey data (kg/km) for all shrimps, capelin, smelts, sandfish, herring, and juvenile pollock <20 cm (*www.fakr.noaa.gov/trawl/index.htm*). The catch data for the 10-year running average of central Gulf of Alaska sockeye and pink salmon were taken from Hare et al. (1999). Sockeye and pink account for up to 90% of the total annual salmon landings in the region.

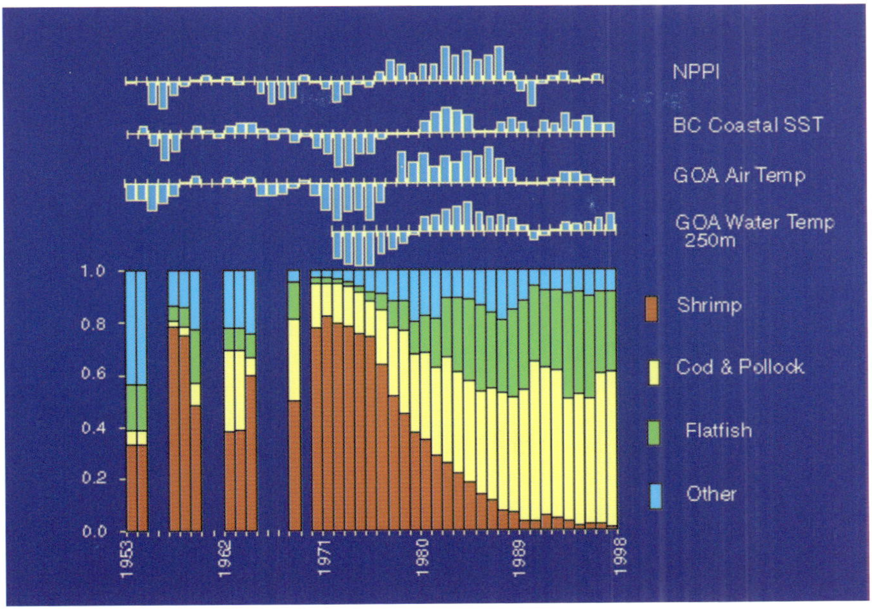


FIGURE 2.4 Proportional species distribution in small-mesh trawl catches in the Gulf of Alaska between 1953 and 1997 relative to climate indices. Climate data expressed as normalized anomalies. NPPI, North Pacific Pressure Index; BC SST, British Columbia sea surface temperature from lighthouse stations; trends were smoothed by taking 3-year running averages. SOURCE: Reproduced from Anderson and Piatt (1999).

## Changes in Bering Sea Biology

The Bering Sea also reflects large-scale circulation and regime shifts, although not necessarily the timing and extent of changes observed in the Gulf of Alaska and vice versa. There was an observed cooling era in the Bering Sea from 1965 to 1978, the main 1977-1978 regime shift, and a warming period from 1979 to 1989 (Wooster and Hollowed, 1995). Niebauer (1998), using sea ice as an indicator of Bering Sea condition, found a significant decrease in ice cover from 1977 into the late 1980s. Recent studies indicate a trend toward earlier retreat of ice in the spring during the 1990s (Stabeno and Overland, 2001), which may have a dramatic impact on ice edge production and associated ecosystem response in the spring (Hunt et al., 2002). Coincidently, over the past 30 years, the Bering Sea has exhibited substantial changes in both pelagic and benthic community structure (National Research Council, 1996). Major environmental changes in the Bering Sea include increased water temperature, possibly influencing the observed increase in zooplankton biomass (Brodeur and Ware, 1992; Sugimoto and Tadokoro, 1997; Brodeur et al., 1999, 2002) invasion by warm-water organisms and a 10-fold increase in gelatinous zooplankton between 1979 and 1997 (Brodeur et al., 1999), and more frequent blooms of coccolithophorid phytoplankton, which can cause substantial mortality of endotherm predators (Baduini et al., 2001; Lovvorn et al., 2001). Various marine birds and mammals have declined precipitously (see Figure 2.5), with food web changes suggested as the likely cause (Springer, 1998, 2001).

Similar ecosystem-level change has been observed in the northern Bering Sea, with indications of benthic biomass changes south of St. Lawrence Island, in the Chirikov Basin between St. Lawrence Island and Bering Strait, and in the southern Chukchi Sea (Grebmeier and Cooper, 2002). Studies begun in the mid-1980s have shown declines in the biomass (Sirenko and Koltun, 1992; Grebmeier, 1993; Grebmeier and Dunton, 2000) and mean sizes of the dominant bivalves in the area (Grebmeier and Cooper, 2002). Sediment respiration rates, which indicate carbon loading to the seafloor, have also declined since the late 1980s. Seasonal patterns of sediment chlorophyll concentrations show that deposition of carbon in this area is closely tied to the ice-edge spring bloom (Cooper et al., 2002); thus, any change in the timing of ice retreat during the late winter/early spring will likely have a major impact on ecosystem structure. Although commercial fishing may have played a role in the trophic changes in the southern Bering Sea, there is little commercial fishing on the northern shelf, where changes in benthic faunal populations and declines in dominant fauna have occurred, with cascading effects on higher trophic levels (Grebmeier and Dunton, 2000; Grebmeier and Cooper, 2002; Lovvorn et al., in press).

Studies by Wyllie-Echevarria and Wooster (1998) have related the extent and duration of the cold pool on the middle shelf of the Bering Sea with the location of fish stocks. The cold pool (<2° C) occurs subsurface (50 to 100 m) and is remnant winter water that is maintained through the summer. Only after the 1977 regime shift (and reduced ice extent and duration) were pollock larvae observed in the northern Bering Sea, with pollock larvae observed north of the Bering Strait in the Chukchi Sea in 1988. Only Arctic cod (*Boreogadus saida*) occur regularly in the cold pool, whereas the subarctic walleye pollock are variable, moving to the outer continental shelf in cold years (when the cold pool expands) but dispersing across the middle domain in warmer years (when the cold pool is smaller). Recent stock assessments (North Pacific Fishery Management Council, 2001b), based largely on annual bottom trawl surveys of the eastern Bering Sea shelf, are generally consistent with the regime shift

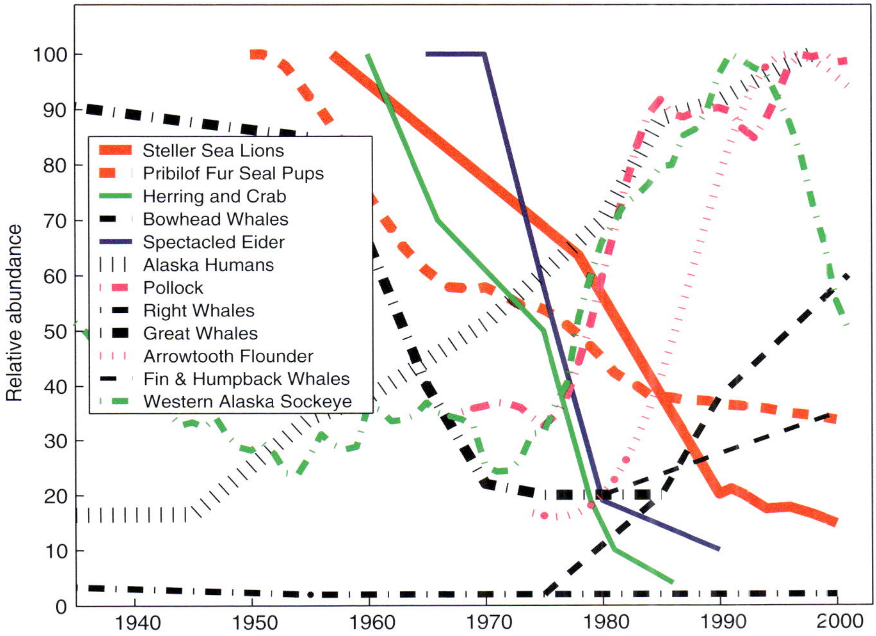


FIGURE 2.5 Bering Sea ecosystem trends: 1945-2001. Schematic diagram indicates the relative abundance of select marine mammals, fish, and shellfish from 1945 to 2001 in the Bering Sea/Aleutian Islands (BSAI). To clarify multiyear trends a 10-year running average was applied to groundfish, forage fish, fur seal pups, and western Alaska sockeye salmon.
SOURCES: Data for blue, fin, and sei and humpback whales (key = Great Whales), and right and bowhead whales are taken from National Research Council (1996). Data from 1996-2002 on bowhead, fin, and humpback whale population trend data came from Robyn Angliss, National Oceanic and Atmospheric Administration, Marine Mammal Lab in Seattle, personal communication, 2002. Data for the northern fur seal annual pup counts on the Pribilofs came from Anne York, National Marine Fisheries Service, Marine Mammal Lab, Seattle, personal communication, 2002. Spectacled eider (*Somateria fischeri*) and herring and crab (except Opilio) data were estimated from National Research Council (1996, Figure 5.8). Steller sea lion, pollock, and arrowtooth flounder data came from National Marine Fisheries Service 2001 stock assessments (*www.refm.noaa.gov/docs/*). Sockeye salmon catch data are taken from Hare et al. (1999). Sockeye account for up to 75% of the total annual salmon landings in this region. Human population came from Alaska census data (*www.labor.state.ak.us/research/pop/popover.htm*).

patterns revealed by Anderson and Piatt (1999) for the Gulf of Alaska. In particular, many demersal fish species (e.g., flatfishes, pollock, cod) increased in the late 1970s to early 1980s in conjunction with declines in some crab species (Zheng and Kruse, 2000), together with anecdotal evidence of declines in shrimp and forage fishes.

# 3

# Identifying Clues and Testing Hypotheses

The challenge in elucidating the causes underlying the initial steep decline and continuing gradual decline in the western population of Steller sea lions (*Eumetopias jubatus*) lies in unraveling the complex multispecies relationships that characterize marine ecosystems. Direct and indirect trophic linkages among the members of these ecosystems present the potential for a population change in one species to profoundly influence the abundance of other species at other trophic levels. There have been several approaches to this problem, including analysis of food webs with few species (Elton, 1927), interaction webs containing a manageable number of multispecies interactions, and ecosystem models. The latter, while incorporating elements of the physical environment, often of necessity aggregate species into functional groupings, thus disguising the ecological roles of individual species (Paine, 1980).

Some blend of the above provides a useful conceptual framework for evaluating various hypotheses concerning the population collapse of Steller sea lions in western Alaska. Both population and ecosystem models are valuable as tools for identifying what types of factors could explain the population decline and for evaluating the available data for consistency with these potential causes. It is important to note, though, that models cannot compensate for the lack of fundamental data for most of the period of decline. Although models can be used to explore potential mechanisms, it will not be possible to establish the causal basis of the sea lion population decline based on these modeling exercises.

## POPULATION MODELS

Population models have been used in conservation biology to estimate population viability, to determine the relative impact of different life history processes on population growth rates, and to predict the effects of management strategies on growth rates. Population viability analysis usually focuses on estimating the probability of extinction over a given time period and is conducted in the early listing stage for endangered species (Groom and Pascual, 1998; Gerber and VanBlaricom, 2001). This approach is limited, however, when there is a need to know which processes (fecundity, growth, or survival) in the life history of an organism are most likely to constrain recovery. To answer this question, researchers often conduct elasticity analyses (the proportional change in population growth rate due to a proportional change in a vital rate such as survival) or stochastic simulations (testing the effect of random variations in population vital rates across their statistical distribution on variations in population growth rate) to determine how variations in particular processes influence expected population growth rate (Heppell et al., 2000). In addition, these analyses can be used to compare the potential explanatory power of alternative hypotheses for population declines.

Population models vary in their structure from simple, deterministic, age-based matrix models to stochastic, individual-based models. Age-based models assume that all individuals of the same age experience similar dynamics. Individual-based models allow variation among individuals of the same age to influence population dynamics. But data are so limiting for most threatened species that the models are often relatively simple, leaving out features that may be important, such as stochasticity or well-defined geographic variation. Still these models have proven to be a useful way to integrate what we know, to point out unequivocally what we need to know, and to guide managers toward those solutions most likely to contribute to population recovery.

There have been a number of population models developed for Steller sea lions (Pascual and Adkison, 1994; York, 1994; York et al., 1996; Gerber and VanBlaricom, 2001). Each addresses different issues regarding population viability, population trends in the context of local populations, and regional metapopulations. Elasticity analyses of 44 population models based on 50 mammal life tables place Steller sea lions in a larger context. Populations of mammals with life histories like Steller sea lions are about twice as sensitive to changes in adult survival as to changes in juvenile survival (Heppell et al., 2000). Adult survival elasticity is 10-fold higher than elasticity related to fertility. So on first principles, factors that cause adult mortality might be expected to be much more important than those that cause juvenile mortality or that reduce fertility. York's (1994) model suggests that the observed 50% decline in adult sea lions from the mid-1970s to the mid-1980s and the concurrent increase in average age resulted from a 10 to 20% decrease in juvenile (ages 0 to 3) survival coupled with a very small decline in adult survival. Hence, the shift in age structure suggests that juvenile sea lions were disproportionately affected by whatever factor (or factors) was responsible for the steep population decline.

Pascual and Adkison (1994) used York's model to examine a series of alternative hypotheses to explain the decline in sea lions. They argued that deterministic transient population behavior (time-delayed population response to an earlier event), historical pup harvesting, and short-term environmental stochasticity are unlikely to explain the decline. They concluded that from 1975 to 1990 sea lion declines were likely caused by a long-term or catastrophic change in environmental conditions that produced a 30 to 60% reduction in juvenile survival or a 70 to 100% reduction in fecundity because of food limitation. Both Pascual and Adkison and York et al. (1996) suggest that these processes vary spatially, leading to different local population trends. Pascual and Adkison acknowledge that anthropogenic causes (competition with fisheries, shooting, incidental takes) or long-term climatic fluctuations could explain the decline through a combination of bottom-up and top-down mechanisms.

One of the most useful applications of population models has been to calculate the unexplained mortality that must have occurred to explain observed census results, assuming that reproductive rate has not decreased over the years. Unexplained mortality refers to the additional number of deaths required to make up the difference between the known mortality and the observed decrease in the population. Figure 3.1 shows one such "reconstruction" of past mortality rates, obtained with a spreadsheet age structure model using age-specific birth and survival rates supplied to the committee by Anne York (National Marine Fisheries Service, Alaska Fisheries Science Center, National Marine Mammal Laboratory, Seattle, personal communication, 2001). The spreadsheet simulated forward from 1950 to the present assuming constant (density-independent) age-specific rate schedules that would result in a steady population in 1950, with additional age-independent mortality rates for the periods shown in the figure. Excel's nonlinear search procedure (Solver) was used to adjust the by-period additional mortality rates so as to match the population trend data as closely as possible.

In agreement with National Marine Fisheries Service calculations, the committee found that unexplained mortalities must have peaked at nearly 30% of the population, or around 15,000 to 20,000 animals per year, during the period of most rapid decline beginning in the mid-1980s. Current (late 1990s) unexplained mortality is estimated to be about 1,000 animals per year, a mortality rate 5% higher than expected for a steady population

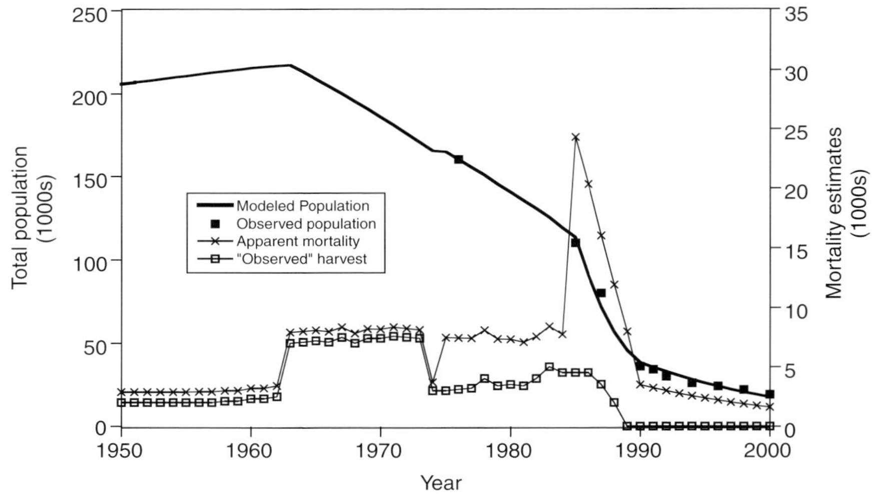


FIGURE 3.1 Fit of an age-structured population model to the western stock abundance data for Steller sea lions, using time-varying estimates of harvest compiled by Trites and Larkin (1992) and fitted by varying time-dependent "unexplained" mortality rates. Age-specific birth and survival rates were assumed to be constant over the simulation period. There was a brief period of very high, unexplained losses during the late 1980s but the unexplained loss from the mid-1970s to the mid-1980s is very close to the estimated commercial harvest from the 1960s to 1970s period of legal harvesting.
SOURCE: Figure based on data from Trites and Larkin (1992), supplied in spreadsheet format by T.R. Loughlin and A.E. York, National Marine Fisheries Service, Alaska Fisheries Science Center, Seattle, 2002.

size and 1,500 animals per year by comparison with the increasing eastern stock (see Box 3.1). These calculations of annual unexplained mortalities are not sensitive to alternative assumptions about the age distribution of the mortalities (e.g., concentrated on pups, yearlings, juveniles, or widely distributed over all ages), except that the losses during the 1980s were too large to have involved only pups and/or yearlings. The estimates are also not sensitive to assumptions about possible compensatory improvements in fecundity or early juvenile survival that might have occurred during the decline. If such improvements did occur, the unexplained mortalities would have to be higher.

During the steepest decline in the Steller sea lion population from 1985 to 1990, estimates of groundfishing-related takes were 1,350 to 1,600

**BOX 3.1**
**The Missing 1,500**

There are now approximately 28,000 to 33,000 nonpup Steller sea lions in the western population compared to about 150,000 in 1977. Most estimates of mortality rates assume a stable (neither increasing nor decreasing) population, but if there were no unnatural mortality, the expectation is that the depressed western stock would increase at a given rate, nominally at least the rate seen in southeastern Alaska. Demographic analyses (Loughlin and York, 2000) generated a projection of Steller sea lion mortality for 2001 that assumed the western stock would continue to decline at the rate observed from 1991 to 2000. If the population were stable (birth rate equal to death rate), there would be 4,710 mortalities. At the current rate of decline (5.2%), another 1,715 mortalities are projected, bringing the total mortality to 6,425 animals. Over this same time period, the Steller sea lion population in southeastern Alaska has been increasing at an average annual rate of 1.7%. Assuming this rate of increase is normal for the population when conditions are favorable, the western population would be projected to increase by 563 individuals. Hence, compared to the southeastern Alaska population, the western stock would be short by 2,280 individuals (projected population of recovering stock, 33,680; projected population of declining stock, 31,400). Loughlin and York made a rough estimate of 779 mortalities from human takes and increases in predation by killer whales and sharks due to human activities. This leaves about 1,500 missing animals (unexplained mortalities projected for 2001) relative to a recovering population.

per year (Perez and Loughlin, 1991; Alverson, 1992). Current takes are estimated to be low, about 30 per year (Ferrero et al., 2000). Subsistence takes have been estimated at 350 animals per year in the 1990s (Wolfe and Mishler, 1997; Wolfe and Hutchinson-Scarbrough, 1999). Shooting of Steller sea lions by fishermen was legal prior to 1972 and until 1990 was allowed in order to prevent gear damage. Trites and Larkin (1996) estimate there were 1,200 annual takes in 1985; it may now be as low as 50 to 100 per year. Predation by killer whales (*Orcinus orca*) is unknown but has been estimated at 1,100 to 1,200 per year based on a rough estimate of killer whale abundance (Barrett-Lennard et al., 1995; Loughlin and York, 2000). Shark predation could account for an additional 130 deaths (Loughlin and York, 2000). The total of these various estimated sources of mortality at their maximum values for the period 1985-1989 comes to fewer than 4,500 annual mortalities. This mortality is three- to fourfold lower than estimated from the model as unexplained mortality. The cause (or causes) of this large excess mortality has largely been attributed to nutritional stress, but it should be noted that estimates of killings by humans and natural predators are based on limited data that may greatly underestimate the actual numbers.

## ECOSYSTEM MODELS

There has been much interest in the possibility that ecosystem models might help to refute or support some of the more complex hypotheses for the Steller sea lion decline, by integrating information on past trophic interactions, fishery effects, and physical forcing factors. Models developed for the eastern Bering Sea system have ranged from relatively simple and static "snapshot" models of trophic mass balance using Ecopath (e.g., Trites et al., 1999; *www.ecopath.org*), to complex spatial models based on the Laevastu et al. (1976, 1982) Dynumes III approach (North Pacific Fishery Management Council, 1999). However, these models have generally not been compared to, fitted to, or "tested" against historical data with the same care as would be expected in single-species stock assessment modeling.

To determine whether trophic models can even come close to explaining historical changes, the committee developed an Ecopath/Ecosim model (Walters et al., 1997, 2000) for the historical period 1950-2000. They started with an Ecopath mass balance model for the 1950 abundances and trophic interactions proposed by Trites et al. (1999) for the eastern Bering Sea. Steps in the analysis are shown in the section below. The Ecopath/Ecosim software system makes it easier to enter time series of historical "forcing" data (fishing mortality rates over time, physical regime indicators over time) and ecosystem "output" data (trends in relative abundances, catches, total mortality rates) than to run multiple simulations that compare model predictions to observed output patterns over time. A similar approach has been taken by the National Marine Fisheries Service (Kerim Aydin, National Marine Fisheries Service, personal communication, 2001).

### Applying Ecopath/Ecosim

Four steps are involved in developing an Ecopath/Ecosim model and "testing" it against historical data: (1) establishing an initial system state, (2) defining a set of rules for dynamic change over time, (3) entering historical reference data, and (4) comparing/fitting predictions from the rules and historical "forcing" input patterns to available data.

The first step is to develop a static trophic flow "snapshot" of the ecosystem at one particular point in time, using Ecopath. This involves partitioning system biomass into a set of functional groups/species and providing basic information on these groups. Biomass estimates must be provided for at least some groups. For all groups, estimates are generally provided of production/biomass (actually total mortality rate), food consumption/biomass (from bioenergetics studies), and diet composition. For groups lacking biomass estimates, estimates are provided of the proportion of total mortality attributable to modeled trophic/fishery flows (consumption by other species and fishery catch). This proportion along with calculated total consumption of the group allows back-calculation of biomass. Estimates can also be provided of the rate of biomass change at the reference time point or year. Ecopath then estimates total biomasses and total fluxes along each predator-prey linkage for a single reference year, so as to account for (balance) all components of biomass change over that year. These estimates provide an initial state for temporal simulations, and the flux components (total biomasses consumed of each prey by each predator in the food web) provide part of the information needed to predict changes in trophic flows over time. The committee used a 24-pool Ecopath model developed by Trites et al. (1999), with one extra pool added to represent the division of small pelagic fishes between nearshore/shallow habitat species like herring (*Clupea pallasi*) and sand lance (*Ammodytes hexapterus*) that are most available to Steller sea lions versus offshore/deepwater species (such as myctophids), which may be available mainly to the benthic piscivores like arrowtooth flounder (*Atheresthes stomias*) that have prospered in the eastern Bering Sea in the past two decades.

Second, Ecosim uses the initial or base state rate estimates from Ecopath and adds additional parameters to predict changes in the trophic interaction (predation) rates for other system states (biomasses) besides the original Ecopath state. That is, Ecosim postulates functional relationships based on the recent behavioral ecology theory of foraging arenas (see Walters and Kitchell, 2001) to predict effects of changes in prey and predator abundances on the predation rate components. The functional models for rate change can include effects of predator satiation, changes in foraging times by prey and predator, and changes in physical factors that affect production rates (in case of primary producers) or prey availability/predator effective search rates. Additionally, Ecosim allows replacement of simple biomass dynamics rate equations with more complex delay difference accounting for population age-size structure for species that have strong habitat/trophic ontogeny.

Third, a simple Excel file is used to define a historical reference period (1950-2000) and to specify annual time forcing historical "inputs" and measured model-testing abundance and mortality "output" patterns over the reference period. The committee used results from single-species assessments to provide fishing mortality rate histories for herring, walleye pollock, cod, flatfish (rock sole, arrowtooth flounders), whales, and

Steller sea lions (culling rates). As outputs, relative abundance series were provided, ranging from the Steller sea lion western stock total abundance estimates to total pollock, cod, and flatfish abundance estimates from single-species stock assessments. Time series estimates of total jellyfish, plankton, and forage fish abundance also were included from Brodeur et al. (1999).

Fourth, Ecosim simulations were used to determine how well the model performs at "replaying" past system changes in two testing improvement modes: exploratory game playing and formal fitting procedures. The exploratory/game playing mode of model fitting tends to lead in one or a few promising directions of explanation but certainly does not rule out others. In this mode, simulated and observed abundance trajectories are compared and obvious model data discrepancies are corrected by changing model parameters that may have caused the discrepancies. For example, in this mode it was quickly found that Ecosim would not predict as much variation as seen in the historical eastern Bering Sea data unless parameters representing vulnerability to predation were increased considerably (for all trophic linkages) from their Ecosim default settings. It was also found that inverse patterns of abundance change for benthic piscivores (increase) versus Steller sea lions (decrease) could not be produced when strong feeding preference for small pelagics was assumed, unless it was admitted that different components of the pelagic biomass might be available to these predator types.

The second mode is a more formal "fitting" of the model by nonlinear search procedures, allowing predation vulnerabilities, initial abundances, and annual primary production "anomalies" to vary. The fitting criterion used was a likelihood function for the observed relative abundance time series, assuming independent and log-normal observation errors for these relative abundances. The formal fitting procedures can demonstrate whether or not the data are consistent to at least some degree with a hypothesis of long-term variation in primary productivity, but this consistency is in no way proof that such variations actually took place. This mode was used only to estimate patterns of apparent change in primary production; patterns that the estimation procedure will "see" in this case are those that simultaneously improve fit to multiple relative abundance time series (i.e., that represent apparent ecosystem-scale productivity anomalies).

As a word of warning, it must be emphasized that step four cannot in principle be used to demonstrate that there is some unique "best" model to describe the historical data. We are dealing with fragmentary, incomplete data, and a natural dynamics that historically involved a multitude of dynamic parameters, most of which have not been measured directly. In such situations there is certain to be a wide variety of model structures and rate parameter combinations that will fit the data equally well (i.e., that are equally good hypotheses to explain the data).

### Ecosim Modeling Applied to the Eastern Bering Sea

The available data are insufficient to develop models that will provide unequivocal conclusions about cause and effect with regard to the Steller sea lion decline. Instead, modeling is used here to identify more plausible scenarios for the decline of the Steller sea lion population.

It is quite easy to obtain reasonably good fits to at least qualitative trends in the abundance of major species groups, especially when Ecosim is allowed to estimate apparent temporal anomalies in annual primary production rates (see Figures 3.2 and 3.3). The apparent anomalies that give best fits (and explain about 30% of the variation in the data series shown in Figure 3.2) appear to involve the sort of decadal regime shifts that have been proposed for the region based on climatic data and anomaly patterns in recruitment rates for single-species stock assessment models. Interestingly, little of the variation is explained by historical changes in fishing mortality rates alone. In terms of trophic interactions, much of the good fit is due to effects following from reduction in sperm and baleen whale abundances, represented as a temporal cascade of events.

The model predicts (hindcasts) that reduction in sperm whales, in particular, likely led to substantial increases in their main prey, squids, and this resulted in considerable increase in squid predation on small pelagics. Reduction in small pelagics was then partly responsible for an initial simulated decline in Steller sea lion population and to the beginnings of an increase in jellyfish (more zooplankton available following reduction in pelagic fish abundance). Pelagic fisheries (herring fishery) then hastened the reduction in availability of prey to Steller sea lions, and permitted increase in other pelagics that in turn fueled an increase in benthic piscivores like arrowtooth flounder. Obviously, each of the steps cannot be tested in this complex causal "hypothesis" with the fragmentary data available on the history of the eastern Bering Sea.

However, the committee was unable to find Ecosim parameter combinations that would predict a short period of rapid decline during the late 1970s or early 1980s for Steller sea lions, related only to changes in trophic interactions caused by measured fishery changes and fitted primary production changes. This is not because the Ecosim functional structure precludes such predictions; see Figure 3.4 for an example where it did predict rapid populaton decline for Hawaiian monk seals (*Monachus schauinslandi*) under combined effects of fishing on lobsters and a climatic regime shift. Note the decline predicted in Figure 3.2 is much more

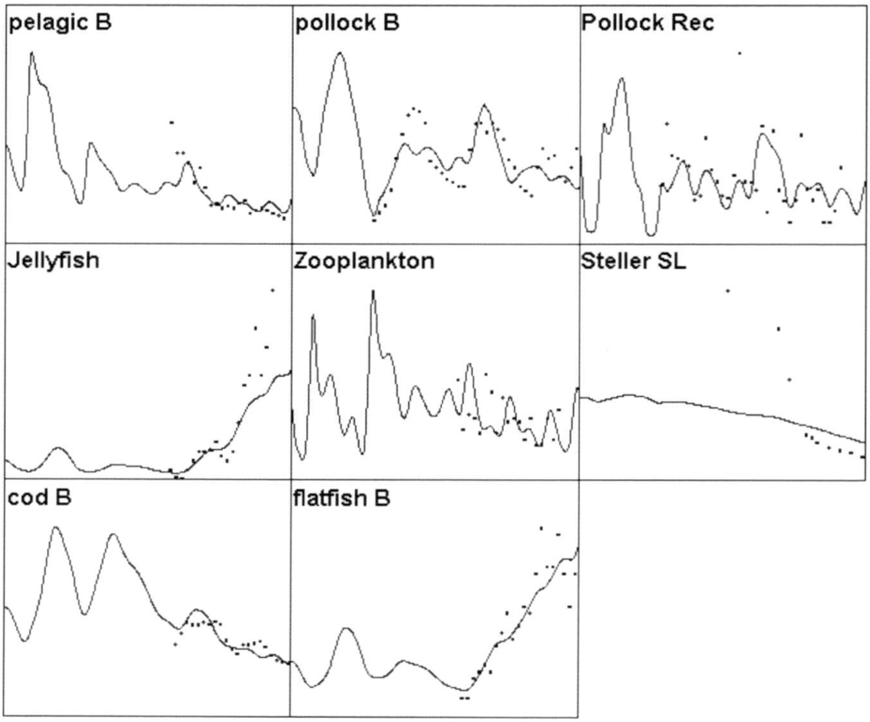


FIGURE 3.2 Ecopath/Ecosim simulation of the eastern Bering Sea ecosystem for the period 1950-2000 compared to selected time series data. Solid lines represent simulated biomasses over time (abscissa is time, 1950-2000; ordinate is relative biomass) and dots represent survey or stock assessment estimates rescaled to the same mean as the simulated biomass series. Simulated biomasses were initialized with the 25 pool Ecopath model based on the mass balance estimates of Trites et al. (1999). Simulated results include effects of time-varying fishing mortality rates, along with fitted estimates of relative primary production rates (fitted relative production rate series shown in Figure 3.3). The base model does not predict as severe or rapid a decline in Steller sea lions as was observed; no parameter combination was found that predicted a rapid decline during the mid-1980s involving only trophic interactions and known impacts of fisheries.
SOURCES: Figure created based on Ecosim simulation of Ecopath model by Trites et al. (1999); times series data for cod, pollock, flatfish came from Blackburn et al. (2001); data on small pelagics, zooplankton, and jellyfish came from Anderson and Piatt (1999).

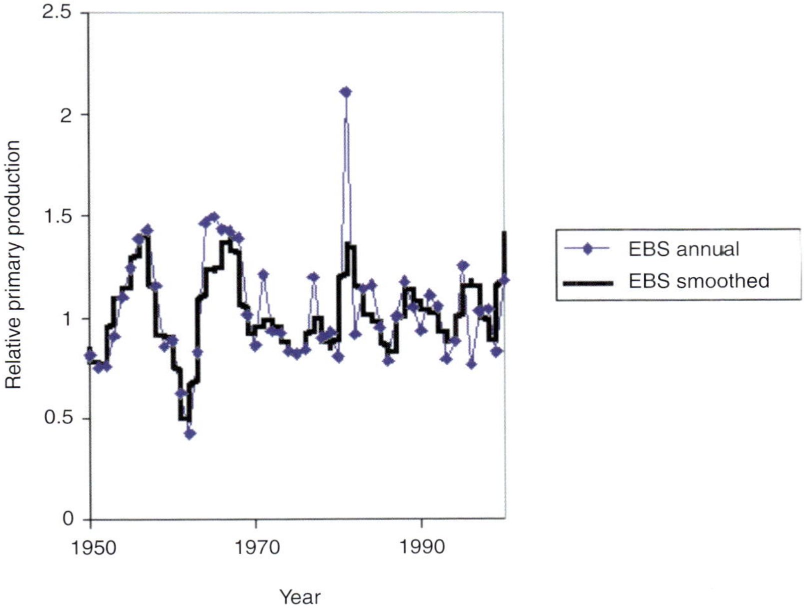


FIGURE 3.3 Fitted time series of relative primary production rates for the eastern Bering Sea (EBS) used to obtain the fits to Ecosim shown in Figure 3.2. A nonlinear search procedure in Ecosim was used to vary annual (or spline function with 2-year nodes) primary production "anomalies" so as to improve the fit of the model to the time series shown in Figure 3.2. These anomaly sequences explain about 30% of the variability in the time series data not explained by historical fishing and trophic interaction effects.

gradual and prolonged than indicted by the census data and that there are no measured rapid changes in prey abundance just prior to or during the rapid Steller sea lion decline. Of course, a more rapid ecosystem-scale regime shift could be forced on the model dynamics in order to produce a more rapid decline in Steller sea lions, but such scenarios result in poor fits to trend data for the fish species.

Good fits to the Steller sea lion data, including the rapid decline, could be obtained by forcing the Ecosim model with various "hidden variable" hypotheses for historical change not directly supported (or refuted) by available field data (see Figure 3.5). One parsimonious hypothesis that gives a good fit (see Figure 3.6) is to simply assume there was much more unreported culling of Steller sea lions by trawl and crab fishers as these fisheries developed and peaked in the early 1980s. Another is

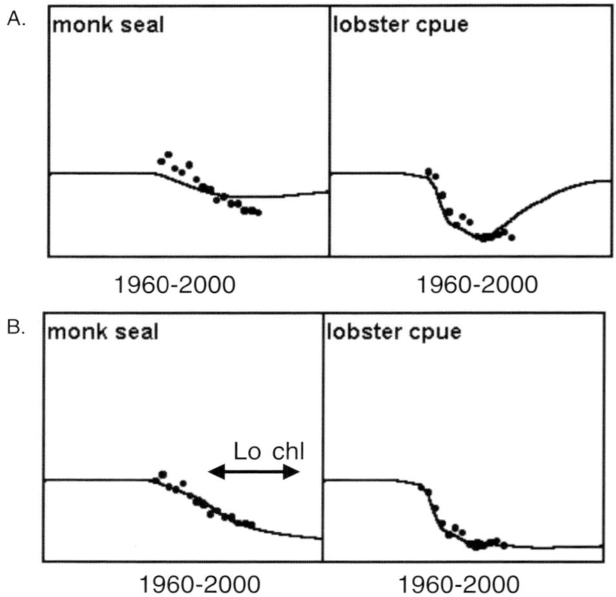


FIGURE 3.4 Prediction of a strong decline in a marine mammal population using Ecosim. Results are for a model of reef ecosystems in the northwestern Hawaiian Islands (French Frigate Shoals), testing whether rock lobster fishery development alone could explain monk seal decline. Data points are for the years 1983-1998. Initial model runs (A) showed that fishing and trophic interactions alone did not explain the monk seal decline and predicted recovery of the lobster population. Polovina's insight (B) was that satellite chlorophyll data indicate a persistent 40-50% decline in primary production beginning around 1990. This example illustrates that Ecosim functional structure is capable of modeling a rapid population decline of marine mammals.
SOURCE: Jeffery Polovina, National Marine Fisheries Service, Honolulu, personal communication, January 1999.

to assume that there were widespread, relatively sudden declines in small pelagics, particularly herring, due to disease outbreaks like those that apparently caused mass mortality in Prince William Sound (Meyers et al., 1994; Stokesbury et al., 2002). Still another is to assume that physical regime changes in 1976-1977 led to reduced vulnerability of various fish prey to Steller sea lions, perhaps by affecting prey depth distributions and/or prey use of coastal areas near rookeries.

There is always the possibility that future, more detailed, modeling

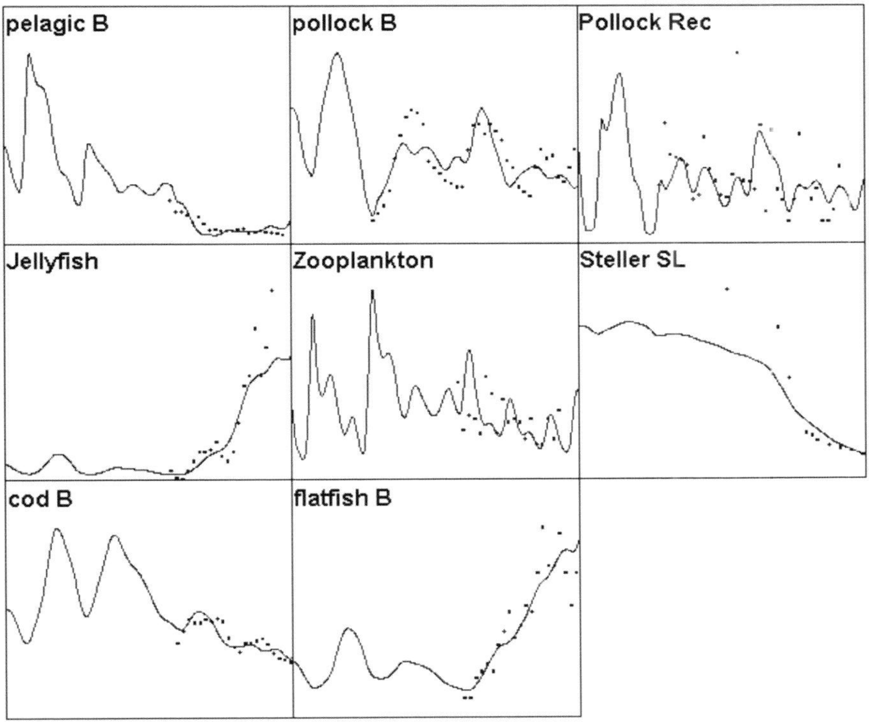


FIGURE 3.5 Comparison of Ecosim-predicted biomass patterns to observed data for the same model as in Figure 3.2 but including a few years of "unexplained" high mortality rate for small pelagic fishes (herring, sand lance) that are available to the Steller sea lion. Note that this scenario gives a somewhat better fit to both the sea lion and pelagic abundance data.
SOURCES: Figure created based on Ecosim simulation of Ecopath model by Trites et al. (1999); times series data for cod, pollock, and flatfish came from Blackburn et al. (2001); data on small pelagics, zooplankton, and jellyfish came from Anderson and Piatt (1999).

exercises, comparison of models for the eastern Bering Sea and Gulf of Alaska regions, and further comparisons to historical data may reveal a "complete" explanation for the Steller sea lion decline in terms of trophic changes. But the committee seriously doubts that this is going to happen, especially in view of gaps in the historical forcing data that would have to be supplied to such models. Hence, models will probably not be sufficient to definitively identify the cause of the rapid decline in the sea lion population.

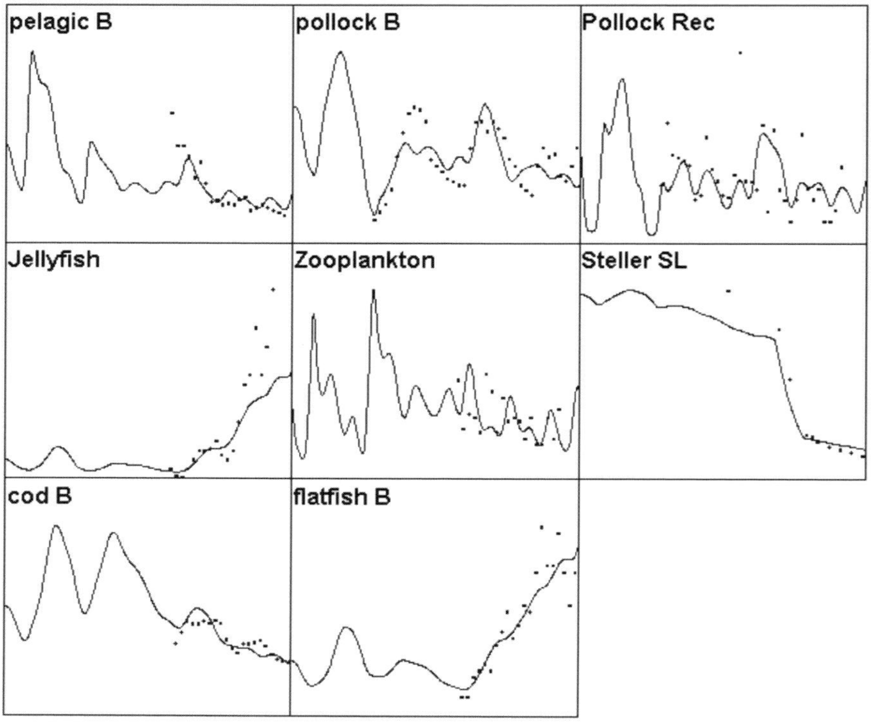


FIGURE 3.6 A further comparison of Ecosim to eastern Bering Sea time series data, but with the addition of direct mortality of Steller sea lions during the mid-1980s due to a hypothesized direct fishery impact (unmonitored culling by crab and pollock fishers). Note that this hypothesis fails to explain the early decline in the pelagic fish abundance index and also fails to explain relatively high rates of decline observed since 1990 in the sea lion population.
SOURCES: Figure created based on Ecosim simulation of Ecopath model by Trites et al., (1999); times series data for cod, pollock, and flatfish came from Blackburn et al. (2001); data on small pelagics, zooplankton, and jellyfish came from Anderson and Piatt (1999).

## EVALUATING HYPOTHESES FOR THE CAUSE OF THE STELLER SEA LION DECLINE

A variety of complex hypotheses involving changes in trophic interactions in the eastern Bering Sea and Gulf of Alaska have been suggested as possible explanations for the Steller sea lion decline. These hypotheses mainly involve factors that could have resulted in two types of immediate effects on the sea lions: (1) reductions in the availability of pelagic, high-quality (fatty), prey like herring and sand lance and/or (2) reductions in alternative prey for predators such as killer whales, which might have led these predators to switch to Steller sea lions. Note that these immediate effects do not involve geographic patterning or directional changes in ecosystem productivity. On a global scale (eastern Bering Sea plus western Gulf of Alaska), the data from single-species stock assessments do not support the proposition that fisheries and/or climatic factors have led to a reduction in the total abundance of food available to Steller sea lions. Some widely distributed and very abundant fish species like pollock and cod have increased considerably over the period of the Steller sea lion decline, as have some bird species (Dragoo et al., 2000). One climatic index, mean wind speed over the eastern Bering Sea, has declined considerably. This could result in a decrease in annual primary production, due to decreased vertical mixing and supply of nutrients to the photic zone. However, the mean wind decrease is not well correlated with most indicators of productivity at higher trophic levels.

Pelagic prey availability could have changed for at least two reasons: changes in prey distributions related to physical changes (e.g., changes in depth distributions associated with warming) and changes in total prey abundance. There is little evidence of major changes in prey vertical or horizontal distributions, but there is considerable evidence of changes in pelagic prey abundance, especially herring. Hypotheses that could explain pelagic species abundance change include:

**1.** ***Herring reduction fisheries (ended in late 1960s) and later roe fisheries.*** Herring stock assessments do not indicate complete recovery in stock sizes following closure of the reduction fisheries as was observed farther south in British Columbia. BC assessments generally show dramatic increases in herring stocks during the early 1970s (Hourston and Haegele, 1980), while Wespestad's (1991) assessment and more recent trends reported by Brodeur et al. (1999) show continuing low abundance in the eastern Bering Sea. However, under this hypothesis, the Steller sea lion decline should have occurred much sooner than currently estimated.

**2.** ***Food web impacts of whale fisheries.*** Whaling in the 1950s and 1960s led to considerable reduction in sperm whale populations (National Research Council, 1996). This may have permitted squid, their main prey, to increase. Squid feed on small pelagics and may now "capture" much more of the net productivity of the pelagics than was possible when whales were abundant. This hypothesis is "supported" by Ecopath/Ecosim modeling results but again should have led to an earlier sea lion decline than was observed.

**3.** ***Competition with jellyfish.*** Brodeur et al. (1999) suggest that major increases in jellyfish abundance since at least 1980 may have resulted in both higher direct predation on eggs and larvae and greater food competition with pelagic, plankton-feeding fishes. However, the rapid decline in Steller sea lions that began in the late 1970s cannot be explained by an increase in jellyfish during the 1980s.

**4.** ***Shift in fish community dominance from pelagic to "benthic" species.*** Herring and other small pelagics may compete with juveniles of pollock and other large fish species for food and may also prey on juveniles of these species. Thus, they are not simply prey for the larger fish. Such reciprocal mortality interactions, where the prey can cause increased juvenile mortality of its predators as well as predators causing higher prey mortality, can lead to so-called cultivation-depensation effects (Walters and Kitchell, 2001) and to multiple equilibria in community structure. In this case, one equilibrium might involve dominance by small pelagics, and another might involve dominance by larger fishes. The shift toward benthic dominance may have been triggered by herring fisheries, climatic regime shifts, or both (Anderson and Piatt, 1999). The benthic piscivore community has increased considerably in biomass despite the coincident development of the fishery (North Pacific Fishery Management Council, 2001a, 2001b). A fishery closure might result in even higher fish biomass absent effects of fishing on piscivore age-size structure, unless cannibalism were to increase considerably. The shifts in community dominance do not coincide precisely with the timing of the Steller sea lion decline; the shifts apparently began during the 1970s and were largely complete by the time of rapid decline in the late 1980s.

**5.** ***Changes in the seasonal timing of water temperature and plankton production.*** There are apparently at least two types of "growing seasons" in the Bering Sea. One involves late ice cover that results in a spring bloom when the water temperature is low enough to inhibit zooplankton productivity. The other involves less ice cover, resulting in a delayed spring bloom with warmer temperatures that favor the growth of zooplankton. The second of these regimes appears to favor recruitment of planktivorous fish and their predators, such as pollock. During a warm regime, strong year classes of pollock could suppress the populations of forage fish (also see hypothesis 4). However, in the Gulf of Alaska, where sea ice is not a factor in regulating the timing of the spring bloom, this mechanism would not apply.

## EVALUATING MECHANISMS IN RELATION TO SEA LION POPULATION DYNAMICS

Considering the natural complexity of the eastern Bering Sea food web, further complicated by the effects of fisheries and oceanographic conditions on the recruitment of individual species, a variety of other hypotheses could be proposed that invoke changes in food web structure. And doubtless, at least some support for these alternative hypotheses could be found in the biological and physical data that have been collected. But all of these stories would share two key features: they require many contingencies to fit the observations, and they do not predict the particular timing of the Steller sea lion decline.

There is good evidence (from fine-mesh trawl surveys) of a rapid shift from pelagic to benthic community dominance in the western Gulf of Alaska at just about the right time to explain rapid sea lion declines in the region from Prince William Sound to the eastern Aleutians (Anderson and Piatt, 1999). Anderson and Piatt argue that this shift may have involved a change in seasonal timing of plankton production (hypothesis 5 above), from a late-bloom pattern that may favor small pelagics, shrimp, and crabs to an earlier bloom pattern that may favor pollock and some flatfishes. They note that increases in benthic predator recruitment would then exaggerate the decline in the smaller, late-spawning species, through top-down predation effects.

It is quite possible that the Gulf of Alaska "community reorganization" involved strong trophic interaction (cultivation/depensation) effects along with the physical regime effects described by Anderson and Piatt. If so, fishery reductions and closures aimed at the groundfish stocks may help to reinforce benthic community dominance and prevent recovery of a community structure that favors Steller sea lions.

Shifts in food web organization due to both fishing and environmental change might have altered feeding patterns of killer whales and other predators, resulting in higher predation rates than might be expected based on a straightforward change in relative prey abundance. Such "predator switching" effects constitute a distinct hypothesis for the Steller sea lion decline, discussed in more detail in Chapter 6. Here we simply note that any of the mechanisms above that may have led to a change in the "prey field" available to the Steller sea lion also may have affected killer whales and other potential predators. These predators may then have compensated by switching prey and exaggerating the effects of the other mechanisms. Ecosim/Ecopath models fit this hypothesis if there is at least a doubling of killer whale predation on Steller sea lions coincident with the Steller sea lion decline and another mechanism (like disease or a cultivation/depensation shift) caused a change in pelagic prey availability to both Steller sea lions and killer whales during the early 1980s.

## SUMMARY

Both the population and ecosystem models demonstrate the necessity of quantitative historical data. Lack of this information constrains the interpretation of modeling exercises, and without input of archival or unanalyzed data, these models will not be able to definitively test any of the current hypotheses about the cause or causes of the historical decline of Steller sea lions. However, these models clearly identify what types of data should be collected to monitor current population trends and sources of mortality.

The models encourage serious examination of many of the previously underemphasized top-down hypotheses. The population models suggest that increased adult mortality more readily accounts for the rapid decline in the 1980s than either reduced fertility or increased juvenile mortality due to food limitation. Therefore, the question shifts away from "Is it food?" to "Were they food?" (e.g., sources of sea lion mortality from predation by humans or other top carnivores). However, these conclusions are based on old demographic data of limited sample size, and until more current data become available, this interpretation rests on a shaky foundation. Nonetheless, population models examined in a global spatial context suggest that multiple factors have contributed to the decline. It is not sufficient to consider regime shifts or extreme fluctuations in availability of important prey populations in the absence of other indirect effects arising from the multiple linkages in food webs. The ecosystem models develop a varied menu of hypotheses to explain the historical decline. Thus, based on available trajectories for the population trends of many species, Ecosim is unable to reproduce the rapid decline in sea lion abundance. The model could not reproduce the steep population decline involving only trophic interactions and known impacts of fisheries. Adding other, often undocumented, mortality factors substantially improves the fit. The value of inventing and using undocumented levels of historical forcing, again acknowledging the constraints imposed by insufficient information, is that they identify a tantalizing variety of possible causes of the decline. These approaches, despite their implicit and identified limitations, provide a framework for evaluating the various hypothesized causes of the continuing decline in the sea lion population and identifying the most fruitful avenues for future research.

# 4

# Review of Steller Sea Lion Biology

## BIOGEOGRAPHY AND EVOLUTION

There are three extant families of pinnipeds—the otariids (sea lions and fur seals), phocids (true seals), and odobenids (walruses). Steller sea lions are members of the Otariidae family. The otariids arose from arctoid carnivores during the Oligocene, presumably by way of an ursid (bear)-like ancestor (Romer, 1966). The otariids themselves originated during the Miocene in the northeastern Pacific Ocean (Repenning et al., 1979), after which they diversified rapidly. Late Cenozoic polar cooling, and the resulting increase in marine production from cold, nutrient-rich seas, likely drove these radiations and many others.

In contrast with phocids, which range from the tropics to polar regions of the Arctic and Antarctica, otariids are generally restricted to temperate latitudes (King, 1983). Hence, unlike phocids, which successfully managed both transarctic and transtropical interchanges, early otariids were confined to the North Pacific Ocean by continental land masses in the east and west, polar sea ice in the north, and the tropics in the south (Repenning, 1976). They remained in this region until Pliocene closure of the Central American Seaway produced cool upwelled conditions in the eastern tropical Pacific Ocean, thus providing an avenue of dispersal along the west coast of the Americas to the southern hemisphere. From southern South America, and aided by the westwind drift, the otariids spread eastward to southern Africa, Australia, New Zealand, and many of the subarctic islands (Repenning et al., 1979). Fur seals first crossed the tropical Pacific into the southern hemisphere some 5 million years ago; sea lions followed about 2 million years later (Wynen et al., 2001). Otariids have never occupied the North Atlantic Ocean, apparently because they were unable to breach the tropical Atlantic.

Traditionally the otariids have been separated into two subfamilies—the fur seals and sea lions. However, this classification is not supported by recent molecular phylogenetic analyses (Wynen et al., 2001). These analyses indicate that the northern fur seal (*Callorhinus ursinus*) has a basal relationship, diverging from a lineage leading to the remaining otariid species about 6 million years ago. Precise relationships among these species are unclear because various clades radiated rapidly at the same time they were diverging from one another. In particular, it is unclear whether the modern sea lions are mono- or multiphyletic.

Seventeen species of otariids are recognized (Gentry, 2002)—10 fur seals (2 genera: *Arctocephalus* and *Callorhinus*) and 7 sea lions (5 genera: *Eumetopias, Neophoca, Otaria, Phocarctos,* and *Zalophus*). Two species (*Zalophus wollebaeki* and *Arctocephalus galapogoensis*) are equatorial, occurring in the upwelled waters of the Galapagos; five species occur in the northern hemisphere (*Z. californianus, Z. japonicus, E. jubata, C. ursinus,* and *A. townsendi*) and the remaining 10 species occur in the southern hemisphere. Otariids vary greatly in overall abundance, from the comparatively rare (more than 7,000 individuals) Guadalupe fur seal (*Arctocephalus townsendi*) to the extremely abundant (about 3 million) Antarctic fur seal (*Arctocephalus gazella*). According to Gentry, only Steller sea lions and southern sea lions (*Otaria flavescens*) are in overall decline, although northern fur seals in the eastern North Pacific Ocean could be added to this list.

In contrast with the phocids, which display a wide array of foraging behaviors and mating systems (Riedman, 1990), both of these features are more highly conserved in the otariids. All species of otariids are sexually dimorphic, with males being larger than females, and all species have polygynous mating systems in which males aggressively defend breeding areas or groups of females. This extreme sexual dimorphism and polygyny probably arose from the combined selective effects of (1) high population densities in productive temperate seas; (2) a dependency on land for birth and reproduction; (3) the limited availability of suitable land-based habitats (e.g., predator-free islands) at temperate latitudes; (4) the resulting aggregation of individuals on land; and (5) the consequent ability of males to compete for and acquire large numbers of females (Bartholomew, 1970). The postpartum behavior of adult females is also highly stereotyped—during this period, all otariid species alternate between at-sea foraging and on-land suckling until their pups are weaned.

## POPULATION STRUCTURE OF STELLER SEA LIONS

Steller sea lions are distributed throughout most of the North Pacific from the coast of California to eastern Russia and northern Japan. Major rookeries and haulouts in the Alaskan part of the Steller sea lion range are shown in Figure 4.1. As indicated in Box 4.1, Steller sea lions are polygynous: males come ashore to establish breeding territories and guard harems of females. Immature animals occupy haulout sites, usually in the vicinity of the breeding and pupping rookeries. Thus, the sex and age structure observed on the breeding rookeries does not really represent the sex and age structure of the population. Existing data suggest sex ratios at birth are near 50-50; thus, there is little reason to believe any bias exists in population sex ratios except possibly in older age groups when males may show higher mortality rates than females. Higher mortality rates for older males are suggested because the competition for breeding among males creates wounds that might contribute to shorter lives, but there are very little data on survival rates at older ages.

During the period of rapid decline of Steller sea lions in western Alaska, the population in southeastern Alaska showed a slow but steady increase. This prompted consideration of the western stock as a distinct population segment as defined by the Endangered Species Act. In 1995 the National Marine Fisheries Service (NMFS) proposed listing the western stock as an endangered population separate from the eastern stock. This consideration was based on genetic studies, tagging and branding studies, and telemetry studies (60 *Federal Register* 192, Oct. 4, 1995, pp. 51968-51978).

With respect to behavior at the colony level that supports separation of the two stocks, Raum-Suryan et al. (2002) carried out a complete analysis of branded and tagged individuals from past marking studies. They found that the observed breeding fidelity to the rookery of birth for females branded in 1987-1988 was 67% for the western stock and 81% for females branded in 1994-1995 in the eastern stock. Sighting of marked females showed that they generally remain within 500 km of the natal rookery. In general, some straying occurred, but homing by females to their natal areas was quite strong. Males tended to use more haulout sites and rookeries and to disperse greater distances. However, there was little evidence of exchange of individuals between stocks. No adults were observed breeding in the opposite stock, but some adults of breeding age did move between stocks.

Genetic studies have mostly used mitochondrial DNA (mtDNA) sequences to characterize the stock structure of sea lions throughout their range. The mitochondrial genome is maternally inherited as a discrete unit and can be used to analyze female lineages. The distribution of female

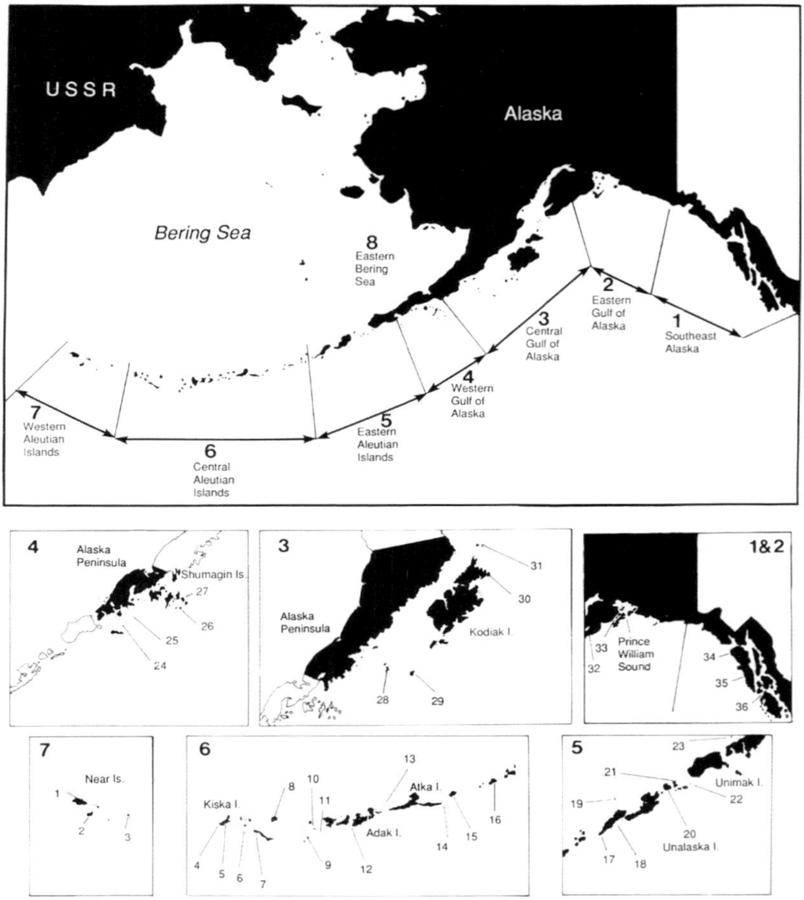


1. Attu
2. Agattu
3. Buldir
4, 5 Kiska
6. Ayugadak
7. Amichitka
8. Semisopochnoi
9. Ulak
10. Tag
11. Gramp Rock
12. Adak
13. Kasatochi
14. Agligadak
15. Seguam
16. Yunaska
17. Adugak
18. Ogchul
19. Bogoslof
20. Akutan
21. Akun
22. Ugamak
23. Sea Lion Rock
24. Clubbing Rocks
25. Pinnacle Rock
26. Chernabura
27. Atkins
28. Chowiet
29. Chirikof
30. Marmot
31. Sugarloaf
32. Outer
33. Seal Rocks
34. White Sisters
35. Hazy
36. Forrester

FIGURE 4.1 Distribution of the Steller sea lion in Alaska. The eight major geographic divisions of the population are shown on the map of Alaska above. The locations of individual rookeries and haulouts are indicated on the lower maps of each division and identified in the key by number.
SOURCE: Sease et al. (2001, Figure 1).

**BOX 4.1**
**Steller Sea Lion Life History**

The Steller sea lion (*Eumetopias jubatus*) is the largest otariid species. Adult males and females average 566 and 264 kg in weight, respectively (Calkins and Pitcher, 1982). They are found along the northwest coast of North America and extend to the eastern Russian and Japanese coasts. The Steller sea lion is a gregarious, polygynous species with an annual seasonal reproductive cycle. They form large rookeries during late spring on isolated islands and rocky shorelines throughout their range. Males arrive at rookeries in May to establish and aggressively defend territories. The largest bulls typically control the prime territories and nonbreeding (bachelor) males remain at peripheral sites or haulouts. Females arrive at rookeries soon after the males and give birth to a single pup within a few days (Pitcher and Calkins, 1981). The majority of pupping occurs between mid-May and mid-July, and breeding occurs approximately 10 days postpartum (Pitcher and Calkins, 1981). There is a period of delayed implantation with fetal development beginning in September. Pups usually remain with their mothers for slightly less than a year before weaning, although some will nurse for a second year (Pitcher and Calkins, 1981; Ono et al., 2001). Females become sexually mature at about 4.5 years of age. Males mature at about 7 years of age but usually cannot maintain a territory until they are larger and older, about 9 to 13 years of age.

SOURCE: Harmon (2001).

lineages throughout the population range provides an indication of the degree of female migration. The mtDNA analysis identified a distinct break in genetic haplotypes between the western and eastern stocks, defined as populations separated at Cape Suckling (144° W; Bickham et al., 1998). Hence, the mtDNA haplotype distribution confirmed earlier tagging studies that showed females tend not to migrate but return to their site of birth to breed (natal homing). NMFS concluded that natal site fidelity made it unlikely that declining rookeries in the western stock would be supplemented by recruitment from more distant healthy populations in southeastern Alaska.

Existing evidence from microsatellite DNA supports the previous tagging data, suggesting that males do not have as high site fidelity as females. Examination of microsatellite DNA markers,[1] found in the nuclear

[1]Microsatellites are short repeated sequences of two to four nucleotide bases that show high variation in many loci of the nuclear genome. It is the most sensitive method to date for measuring genetic variation among populations.

genome, shows less genetic differentiation of eastern and western stocks than the mtDNA (Loughlin and York, 2001). Since the nuclear DNA of an individual includes both maternal and paternal genes, this indicates that males have migrated among the rookeries over larger distances than the females.

## POPULATION TRENDS

Counts of Steller sea lions began as early as 1959 at rookery sites from islands just north of Kodiak Island out the Aleutians to Attu (see Tables 4.1 and 4.2). At this time there was no reason to believe that any decline in the number of sea lions would be possible since it was reported that there were probably over 300,000 in the world. In the late 1970s a group of biologists surveyed the Aleutians via boat and declared the "stock to be healthy and robust" (Loughlin, 1998). Although counts in those early days were not repeated at frequent intervals, they do provide some baseline data that indicate when the western stock of sea lions began its most rapid declines and in what area the declines were initially the most dramatic. There was likely a considerable difference in the quality of the counts between the early days and current counts, mostly because of improved technology and logistics. The early counts were made using several techniques, some from boats moving along the shore, others from shore counts by people on the rookeries, and still others from fixed-wing aerial observations. These counts were considered to be a total tally of individual sea lions observed. At present, the number of sea lions observed in the surveys represents the minimum population estimate and there is no attempt to correct for animals that were at sea when the counts were taken. Therefore, these values represent an index, not an exact count of the number of sea lions at any given site. Techniques have improved over time, and the most recent counts were made using standardized techniques with experienced observers from fixed-wing aircraft. Also, new imaging procedures have enhanced the applications of aerial photography.

The exact timing of declines in all areas of the range has been difficult to establish because frequent (on a schedule of about every 2 years) range-wide counts did not begin until 1989 (Loughlin et al., 1992). Braham et al. (1980) identified a decline in the eastern Aleutians by comparing aerial surveys carried out during the summers of 1975, 1976, and 1977 with surveys conducted between 1957 and 1968. These surveys covered the rookeries and haulout areas in the vicinity of Umnak and Unalaska Islands and documented a decline of 50% in this region between the survey periods. Loughlin et al. (1984) summarized existing data for the entire Steller sea lion range and noted declines in two specific areas between 1956-1960 and 1975-1980 (see Figure 4.2).

TABLE 4.1 Estimates of the Number of Nonpup Steller Sea Lions by Year and Geographical Subarea. Each subarea is characterized by the count data, the fraction of world population and the change in the size of the population from 1960 to 1989.

| Area | 1960 | Percentage[a] | 1977 | Percentage | 1985 | Percentage | 1989[b] | Percentage | Change[c] |
|---|---|---|---|---|---|---|---|---|---|
| Russia | 52,000 | 21 | 29,000[d] | 15 | 18,000[d] | 13 | 13,310 | 15 | –74% |
| Bering Sea (U.S.) | 7,000 | 3 | 2,000 | 1 | 1,000 | <1 | 887 | <1 | –87% |
| Aleutian Islands | 99,000 | 40 | 90,000 | 46 | 61,000 | 45 | 19,033 | 22 | –81% |
| Gulf of Alaska | 69,000 | 27 | 55,000 | 28 | 38,000 | 28 | 31,600 | 34 | –54% |
| Southeastern Alaska | 7,000 | 3 | 8,000 | 4 | 8,000 | 6 | 12,303 | 14 | +70% |
| British Columbia | 8,000 | 3 | 5,000 | 3 | 5,000 | 4 | 8,131 | 9 | 00% |
| Oregon and California | 8,000 | 3 | 5,000 | 3 | 4,000 | 3 | 5,357 | 6 | –33% |
| Total | 250,000 | | 194,000 | | 135,000 | | 90,621 | | –64% |

[a]Percentage of the total for that year.

[b]The counts obtained in 1989 multiplied by 1.331 (pup correction method), except for British Columbia for which 1987 was used.

[c]Percentage change from 1960 to 1989.

[d]The 1977 and 1985 values for Russia were calculated assuming the population declined by a constant number of animals each year from 1960 to 1989. Actual values were not available.

SOURCE: Loughlin et al. (1992), Table 8, p. 237.

TABLE 4.2 Population Trends at Rookery Sites. Trends are presented as regression correlation coefficients (r) where a positive value indicates an increasing slope and a negative value a decreasing slope. Distance refers to nautical miles measured from Outer Island off the Kenai Peninsula. This table indicates that the period of most rapid decline at many rookeries occurred between 1985 and 1989. See also Figure 4.3.

| Rookery | Distance (nm) | 1959-1975 | 1976-1985 | 1985-1989 | 1989-1994 | 1991-2000 |
|---|---|---|---|---|---|---|
| Outer | 0 | 0.0158 | −0.1354 | | −0.2070 | |
| Sugarloaf | 64 | −0.0436 | −0.0620 | −0.1186 | −0.1129 | −0.077 |
| Marmot | 83 | 0.0493 | −0.0759 | −0.1899 | −0.1358 | −0.104 |
| Chirikof | 283 | 0.0181 | −0.0021 | −0.1519 | −0.2117 | −0.183 |
| Chowiet | 298 | −0.0579 | 0.0032 | 0.2568 | −0.0523 | −0.053 |
| Atkins | 405 | −0.0303 | −0.0696 | −0.1818 | −0.0446 | −0.034 |
| Chernabura | 420 | −0.0635 | −0.1202 | 0.0277 | 0.0463 | 0.005 |
| Pinnacle Rock | 484 | −0.0310 | −0.0105 | −0.0376 | −0.0682 | −0.024 |
| Clubbing Rocks | 506 | −0.0129 | 0.0031 | −0.0949 | 0.0074 | −0.021 |
| Sea Lion Rocks | 506 | −0.0127 | −0.1776 | −0.1118 | 0.0774 | 0.015 |
| Ugamak | 590 | −0.0494 | −0.1601 | −0.3015 | 0.1096 | |
| Akun | 610 | −0.0077 | −0.1232 | −0.2662 | 0.1234 | 0.007 |
| Akutan | 632 | −0.0347 | −0.1062 | −0.1966 | 0.0909 | −0.041 |
| Bogoslof | 698 | 0.0043 | −0.0741 | −0.1588 | −0.1084 | −0.079 |
| Ogchul | 737 | −0.0508 | −0.0907 | −0.2311 | −0.0117 | −0.060 |
| Adugak | 762 | 0.0184 | −0.0821 | −0.2226 | −0.0435 | −0.046 |
| Yunaska | 812 | 0.1329 | −0.1236 | −0.2080 | −0.0098 | −0.063 |
| Seguam | 882 | 0.0499 | −0.1319 | −0.3966 | −0.0034 | −0.027 |
| Agligadak | 900 | 0.0070 | −0.1098 | −0.3340 | −0.5936 | −0.155 |
| Kasatochi | 978 | 0.0047 | −0.1026 | −0.1435 | −0.1801 | −0.013 |
| Adak/Lake Pt. | 1039 | −0.0496 | 0.0069 | −0.2780 | 0.0955 | 0.030 |
| Gramp Rock | 1086 | 0.0445 | −0.0465 | −0.1366 | −0.0650 | −0.032 |
| Tag | 1092 | 0.0735 | −0.1019 | −0.1175 | −0.1259 | −0.027 |
| Ulak | 1109 | 0.0185 | 0.0382 | −0.2220 | −0.0647 | −0.057 |
| Ayugadak | 1179 | 0.0446 | −0.1224 | −0.1476 | −0.0706 | −0.076 |
| Lief Cove | 1213 | 0.2498 | −0.1768 | −0.3032 | −0.0884 | −0.056 |
| Cape St. Stephen | 1218 | 0.0723 | −0.0814 | −0.2672 | −0.1773 | −0.066 |
| Buldir | 1251 | 0.0349 | | | −0.2196 | −0.126 |
| Agattu | 1324 | 0.0033 | −0.1379 | −0.0777 | −0.0766 | −0.096 |
| Attu | 1354 | 0.0037 | −0.0600 | −0.2239 | −0.0114 | |

SOURCE: Modified from Table 12.2 of York et al. (1996); additional data for 1991-2000 came from Anne York, National Marine Fisheries Service, National Marine Mammal Laboratory, Seattle, personal communication, 2002.

As depicted in Figure 4.2, a rather dramatic decline occurred in the region between about 157° W and 165° W longitude (western Gulf of Alaska to eastern Aleutians) and between 152° W and 153° W (central Gulf of Alaska). The region between 157° W and 165° W corresponds to

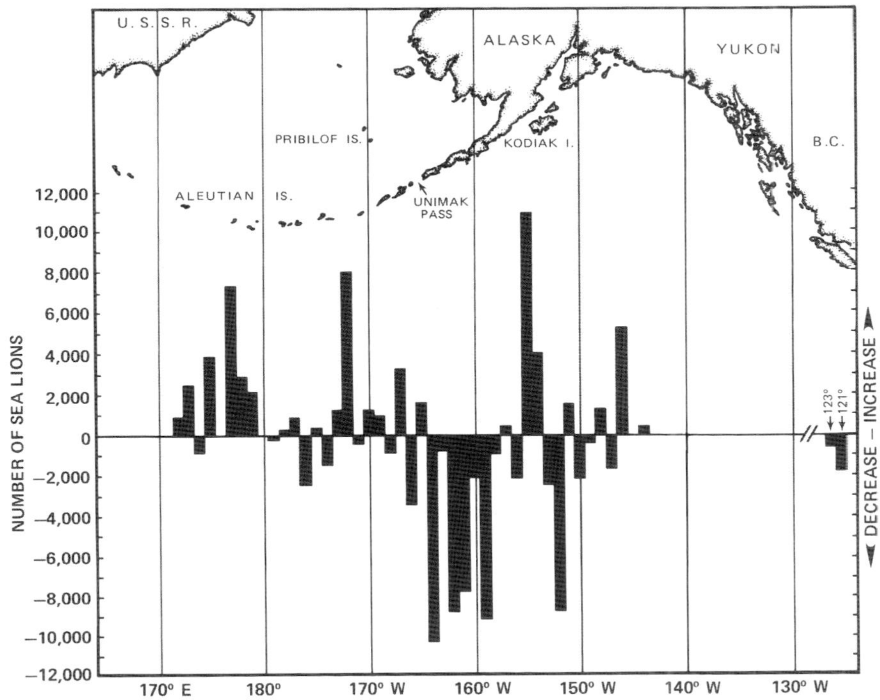


FIGURE 4.2 Changes in site counts (comparison of 1956-1960 and 1975-1980 counts) for which data are available by 1° longitude throughout the central range of Steller sea lions. An upper bar denotes an increase and a lower bar a decrease in the number of Steller sea lions.
SOURCE: Loughlin et al. (1984, Figure 2).

the area where Braham et al. (1980) noted a 50% decline, and the other region corresponds to the population in the vicinity of Kodiak Island. During the late 1980s several publications continued to document declines. Table 4.1, taken from Loughlin et al. (1992), indicates large declines since the 1960s, and the declines are range-wide except in southeastern Alaska.

The homing behavior of female Steller sea lions suggests that rookery sites may represent a metapopulation (Raum-Suryan et al., 2002). A metapopulation is comprised of local populations with low levels of exchange with one another. Each local population may fluctuate or approach extinction in response to local or global conditions. York et al. (1996) considered this possibility and analyzed the count data from the trend rookeries between Kenai and Kiska (see regions 3, 4, 5, and 6 in Figure 4.1). While they were interested in predicting the probability of persistence over time, they summarized data that provide a look at site-specific declines over time. Table 4.2 presents the results of regression analysis on the trend data to calculate rates of change and includes information from more recent censuses. Bar graphs of these data are shown in Figure 4.3. The regression analysis shows that the most dramatic decline occurred between 1985 and 1989 and that this rapid decline included the entire western stock (Table 4.2; Figure 4.2).

The pattern suggests some initial cause(s) that started in the Gulf of Alaska to the eastern Aleutian Islands from Chowiet to Ogchul some time between 1959 and 1975 and spread across the entire western stock by 1989. During the 1989-1994 period, it seems the general area from Chernabura Rocks to Akutan had begun to recover, but then between the 1996 and 2000 censuses this pattern of recovery did not continue. The most recent count data (National Marine Fisheries Service, 2002) showed about a 5% increase in the adult count in the western population above what was found during the 2000 census. This increase should be interpreted cautiously because very few rookeries showed increases in pup counts, and population gains were mostly limited to the Gulf of Alaska with counts in the western Aleutians still in decline. The heterogeneity in population trends since 1990 suggests there may have been a change in the predominant sources of mortality from the earlier to the current pattern of decline.

## VITAL RATES

All Steller sea lion population models use vital rates data derived from samples taken near Marmot Island from 1975 to 1978 and again from 1985 to 1986 (Calkins et al., 1998; Pitcher et al., 1998). Fecundity was estimated by shooting adult females on rookeries, at haulouts, and in coastal Gulf of Alaska (Pitcher and Calkins, 1981; Pitcher et al., 1998). Age estimates were based on bands in the cementum laid down each year of life, and thus the animal's age was estimated by counting these bands from thin sections of the tooth. The second upper premolar was the tooth usually used for sea lions. Pitcher et al. (1998) also measured mass, length, and thickness of blubber, as well as pregnancy and lactation status. Nearly all females were pregnant in both time periods, and heavier females were more likely to be pregnant than lighter females. Standard length, axillary girth, and mass were lower in the 1980s than in the 1970s—most of these differences were manifest in the juvenile stage (Calkins et al., 1998). Sample sizes for the reproductive parameters were 46 and 62 adult females in the 1970s and 1980s, respectively. Age and growth were based on more animals (80 and 102 animals, respectively), but overall sample sizes were not large for either measure. Calkins and Pitcher (1982) used the data on ages of animals shot from 1975 to 1978 to estimate age-specific survival rates for females age 1 to 13+ and adult females collected to estimate fecundity (Pitcher and Calkins, 1981) for a preliminary life table (Calkins and Pitcher, 1982).

It is on these data that all subsequent population modeling has been based—it is unknown whether the relatively small numbers of animals taken were representative of the extant population at the time of sampling. Despite the limitations of the 1970s data, no new vital rates data have been collected since then. Killing animals to obtain new data is unlikely to be permitted given the current population status. Current pup branding protocols allowing long-lasting marks coupled with intense monitoring for resightings should yield critical juvenile survival data. However, it will take some time and a consistent resighting effort to produce reliable new estimates for the older age classes.

Juvenile live captures would give additional information on this vulnerable life stage, and juvenile branding could contribute to survival estimates. No reliable new data on adult survival, at least to ages 10 to 15, will emerge from the pup branding effort until this period of time has passed. Age at first reproduction for females should start to become available after 5 or 6 years. Some data on sexual maturity for branded females have been given by Raum-Suryan et al. (2002), but the sample sizes were small. Significant sample sizes will only become available if sufficient numbers of pups are marked and monitored over long time periods. New age structure data for the present population will be required, possibly from a combination of ages of animals taken in subsistence takes, incidental captures in fisheries, or other sources of mortality.

## PHYSIOLOGY, BEHAVIOR, AND FEEDING ECOLOGY

To understand the potential impacts of food limitation or changes in food quality on the biology of Steller sea lions, it is necessary to consider their physiology, nutritional demands, and behavioral feeding patterns. This is usually conducted by equating their metabolic demands with the amount or quality of food available and the costs of obtaining that food. However, despite the apparent simplicity of this question, obtaining robust data on metabolic expenditures in wild animals and the quality or quantity of food consumed is exceedingly difficult in practice. Some of this is because the animals live in remote and difficult-to-reach locations.

Because nutritional stress is central to many of the hypotheses concerning the decline of Steller sea lions, it is essential to understand the assumptions used in models for evaluating such a stress. If one breaks down the cost of living at sea into the two major components of metabolic

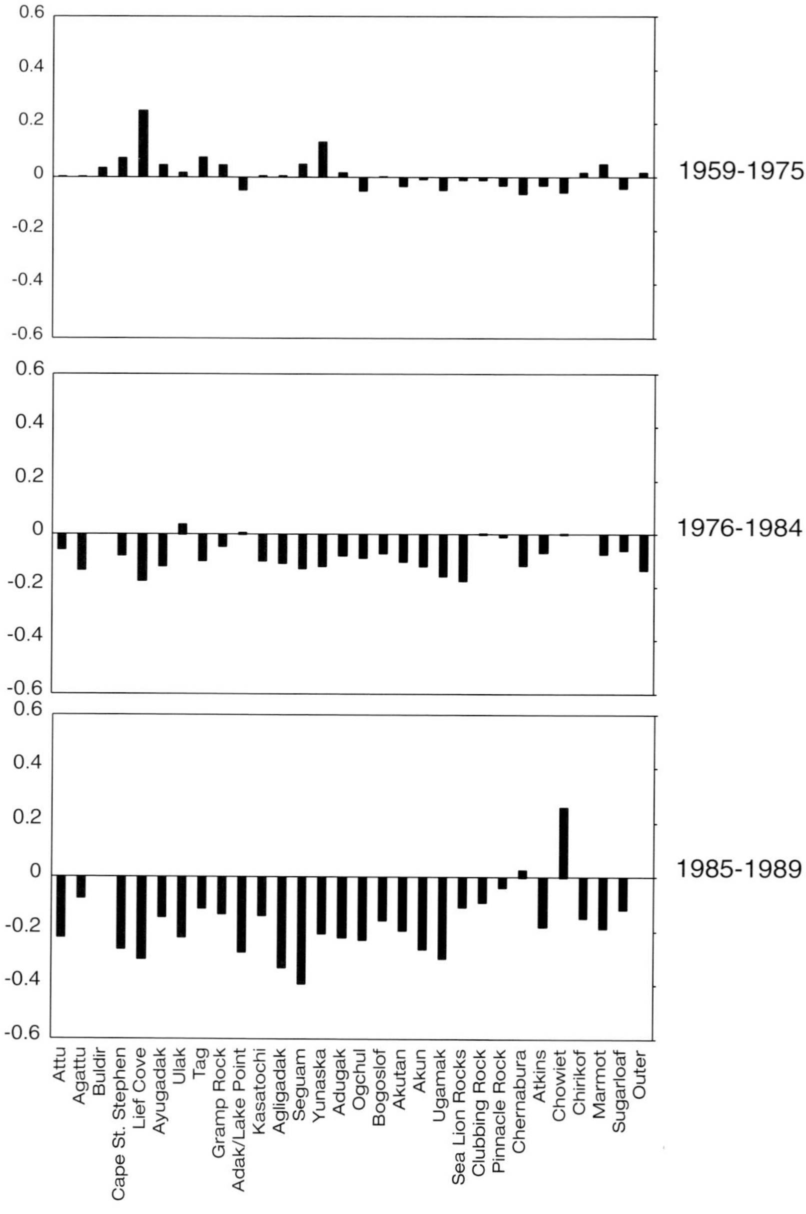


FIGURE 4.3 Continued on next page.

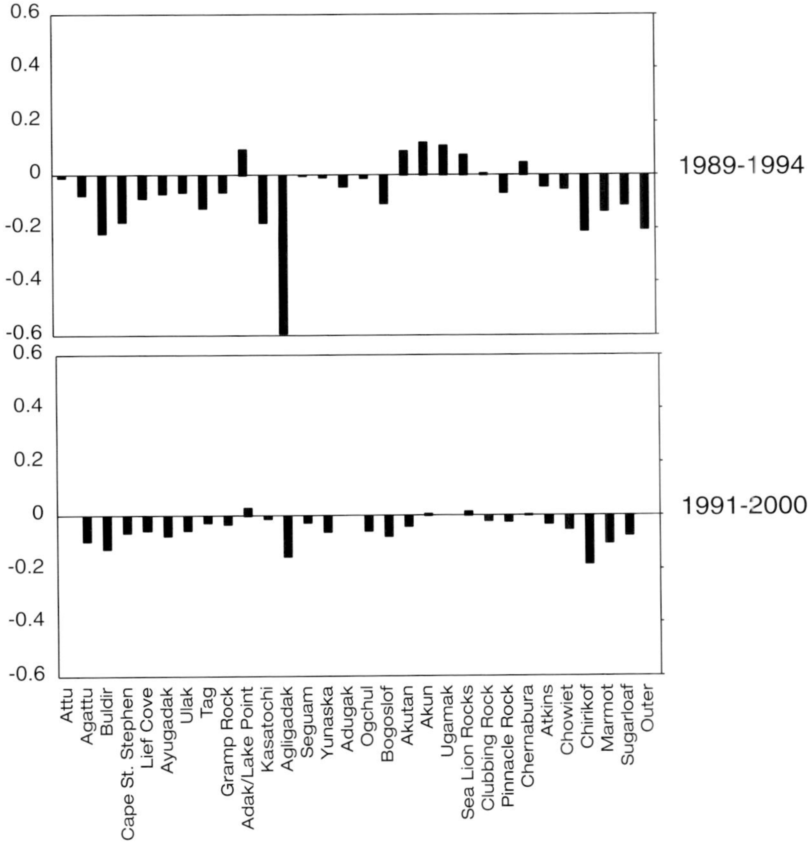


FIGURE 4.3 Relative population trends during various time intervals at rookeries oriented from west to east (Attu to Outer). Values were taken directly from Table 4.2.
SOURCES: Data for 1957-1994 from York et al. (1996); data from 1996-2000 from Sease et al. (2001).

demand and supply, there are distinct biochemical, physiological, and behavioral issues to each of those that can be addressed. Some of these factors can be measured in free-living species but others cannot and must be estimated or modeled. In all cases the goal is to obtain data that can be used to estimate the nutritional requirements of Steller sea lions.

A critical factor in any study of nutrition for pinnipeds is that these animals are well adapted to fasting as a routine part of their annual breeding cycle. Northern elephant seals (*Mirounga angustirostris*), for example, fast from food and water for several months during the breeding season. They metabolize their blubber during this period and use water obtained as a by-product of fat metabolism. They also reduce their metabolic rate in order to minimize metabolic costs. Otariids are fasting adapted, though not for periods as long as most of the seals. Adult female Steller sea lions will fast for about 2 weeks when they come to shore to pup and breed. Adult males will fast for the entire breeding season if they are holding a territory. Therefore, standard mammalian medical, biochemical, and physiological models of starvation do not apply to these fasting-adapted species (Castellini and Rea, 1992). Fasting must be distinguished from starving in Steller sea lions to effectively study food limitation as a cause of population decline. There have been several laboratory studies of fasting in this species that have been designed to describe the body condition and metabolic changes associated with short-term (up to 2-week) periods of food deprivation in Steller sea lions (Rea and Nagy, 2000; Rea et al., 2000, 2001). These fasting-associated changes have been used to estimate feeding status in wild populations as a test of the food limitation hypothesis (Rea et al., 1998).

### Behavior and Feeding Ecology

Steller sea lions consume a variety of fish species and generally dive in the top 250 m to catch their prey. There are developmental aspects to their diving patterns, with young animals diving neither as long nor as deep as older animals (Merrick and Loughlin, 1997). Recent data indicate that young-of-the-year (5 to 6 months old) dive to about 70 m, while pups slightly less than a year old dive in the 140-m range (Rehberg et al., 2001). Routine dive time for young animals remaining underwater is less than 1 minute and for older animals up to about 2 minutes (Rehberg et al., 2001). These data are important for defining critical habitat and understanding if changes in prey base move some fish stocks out of sea lion foraging range. For example, male northern elephant seals migrate into the same area of the North Pacific to feed but do so at depths far exceeding those of routine depths by Steller sea lions (Le Boeuf, 1994). Adult Steller sea lion females with newborn pups will go to sea to feed and return to nurse their pups in 24- to 48-hour cycles. Changes in trip time have been hypothesized to be an indicator of feeding success, as discussed in Chapter 6. Biological Opinion #4 used these types of dive data to prescribe areas of critical habitat and fishing regulation zones.

## STELLER SEA LION DIET

As with most pinniped species, the earliest records of diet for Steller sea lions came from the stomachs of killed animals or from direct observations of animals capturing prey at the surface (Mathisen et al., 1962; Thorsteinson and Lensink, 1962). Given their decline, Steller sea lions can no longer be shot, except by subsistence hunters in certain areas. Current methods of diet analysis include collection and analysis of scat samples from rookeries or haulouts and analysis of chemical signatures in Steller sea lion tissues that are indicative of certain prey items (stable isotope and fatty acid analysis). Methods under development include the use of video dive recorders and measurements of prey ingestion with stomach temperature sensors.

Most of the diet information from stomach collections is for adult animals. Information about the diets of young animals, especially during the weaning transition, is sparser and is based mainly on newer methods using chemical signatures and scat analysis.

Data on the temporal variation in Steller sea lion diets is confined almost exclusively to scat and chemical analysis. The extensive stomach diet data of the 1970s and 1980s was obtained from animals collected during the spring and summer, with few collections made during the haulout periods. There are sporadic notes of unusual diet items for Steller sea lions, including birds, otters, and small pinnipeds, but these are not considered further in this report.

### History of Diet Studies

#### Early Stomach Analysis

Imler and Sarber (1947) collected the stomachs of 23 animals from southeastern Alaska and from the northern Gulf of Alaska (near Kodiak and Seward) in 1945-1946. Of these, 15 (7 from the Gulf of Alaska and 8 from southeastern Alaska) contained food. Pollock was the principal item by volume in 6 of the 8 animals from southeastern Alaska and in 3 of the 7 from the Gulf of Alaska. Salmon were the only fish remains found in 2 animals collected near Seward. Overall, pollock constituted 55% of the volume of fish remains. Wilke and Kenyon (1952) noted that most sea lions collected in the Bering Sea from 1949 to 1951 had empty stomachs. Of 23 animals collected, only 2 had food in their stomachs. One contained only sand lance, and the other had 75% flatfish by volume, 15% cod, and 10% pollock.

In the stomachs of 382 males, primarily harem bulls, from five major rookeries in the western Gulf of Alaska in 1959, Thorsteinson and Lensink

(1962) found that the leading prey by frequency of occurrence were cephalopods (36%), bivalve and gastropod invertebrates (29%), sand lance (18%), and rockfish (11%). Interestingly, they noted no pollock in these stomachs; however, unidentified fish remains were found 18% of the time. In sea lions collected during 1960-1962, Fiscus and Baines (1966) examined the stomach contents of 10 Bering Sea animals and found only one with pollock; the rest had primarily capelin. In 8 animals collected from the Gulf of Alaska, they found salmon, rockfish, capelin, and herring. From 250 animals collected in 1975-1978 in the Gulf, Pitcher (1981) found the dominant food item was pollock (58% of volume, present in two-thirds of the stomachs).

Animals off Oregon were considered "opportunistic feeders, and the stomach contents usually reflected the relative abundance of prey in the area" (Maser et al., 1981). Calkins and Goodwin (1988) analyzed the stomach contents of 170 sea lions collected in southeastern Alaska and around Kodiak Island in 1985-1986. Eighty-eight animals had food in their stomachs (14 from the southeast, 74 from Kodiak). At both sites, pollock was the dominant item by frequency of occurrence (about 60%). Pollock was the largest by volume at Kodiak (42%); unidentified flatfish were second at 25%. In southeastern Alaska, Pacific cod was the largest by volume (57%), and pollock was 32% of the volume. Both studies indicated that the animals fed on a wide variety of fish.

In the proposed designation of Steller sea lion critical habitat, the National Oceanic and Atmospheric Administration summarized the diet studies as follows:

> Data on SSL [Steller sea lion] prey consumption are fairly limited . . . studies in Alaska since 1975 indicate that walleye pollock has been the principal prey in all areas over this time period . . . Few data are available in Alaska prior to 1975 . . . but indicate that pollock may have been a less important component in previous years. (58 *Federal Register* 61, April 1, 1993, p. 17184)

In the 1992 Steller sea lion recovery plan, only stomach content analysis was summarized, with no mention of scat samples. The recovery plan notes that there may be seasonal, spatial, and individual variations in diet but concludes that "diet studies over the last 15 years show that Steller sea lions eat a variety of fishes and invertebrates; demersal and off-bottom schooling fishes predominate." A summary table of stomach content studies can be found in the 1999 Biological Opinion (BiOp #2; Table 5 of National Marine Fisheries Service, 1999) and in condensed form in an overview report prepared by the Alaska Department of Fish & Game (Kruse et al., 2000, Table 3.2).

When concern about Steller sea lions intensified in the early 1990s, information on the types of food and the spatial and temporal variations in their diet became more critical for listing and legal purposes but was harder to obtain because animals could not be taken for stomach content analysis. Therefore, more emphasis has been placed on scat collections and chemical analysis of Steller sea lion tissues. Scats are relatively easy to collect but may not accurately reflect the species composition and abundance of consumed prey because the efficiency of digestion varies with different prey sizes and types. Species composition of scat samples can indicate spatial and temporal patterns, but these require careful interpretation. Large carnivores like sea lions can simply strip flesh from larger prey without consuming any hard parts. Finally, scat analysis cannot be used to assess whether or not a stock is starving or flourishing because it does not reflect caloric consumption. Residence time in the digestive tract, whether a prey species contains hard parts that appear in scats and can be identified, and whether the size and numbers consumed can be inferred from this are essential for determining an energy budget. Thus, scat analyses by themselves have little to offer for resolving issues of food limitation in Steller sea lions.

Chemical analysis of sea lion tissues, such as blubber, provides an indication of what animals have been eating over longer time periods than either stomach contents or scats, but this method is limited because animals must be captured to obtain a sample. Furthermore, chemical analyses (such as stable isotope patterns and fatty acid analysis) are not yet capable of identifying individual prey items in the diet, although ongoing controlled diet studies may make this possible in the future (Iverson, 2001). Development of techniques to obtain blubber samples from dart biopsies, such as used in cetacean studies, could be useful in collecting large numbers of blubber samples without having to capture animals. At present, research on the sea lion's diet is restricted to scat samples and the limited number of tissue samples available for chemical analyses. Bowen et al. (2001) criticize the November 2000 Biological Opinion (BiOp #3; National Marine Fisheries Service, 2000) for concluding that scats are a "reliable tool for monitoring seasonal and temporal trends in predator diets" and recommend greater use of techniques such as fatty acid signatures. However, Bowen et al. do not mention that the relatively small sample size will limit interpretation of the results.

Stable isotope patterns can be used in retrospective studies by utilizing samples from past collections. Hirons et al. (2001) examined bone samples from Steller sea lions, northern fur seals, and harbor seals collected in the Bering Sea and the Gulf of Alaska from 1951 to 1997 for possible changes in stable isotope ratios of nitrogen ($\delta^{15}N$, which varies with trophic level) and carbon ($\delta^{13}C$, which varies with the carbon isotope composition of primary producers). Change in $\delta^{13}C$ might thus reflect shift at the base of the oceanic food web and dietary shifts by these apex predators might be reflected in $\delta^{15}N$ variation. The $\delta^{13}C$ analyses of Steller sea lions suggested that a subtle change might have occurred at the base of the food web. Although similar patterns were not found in harbor seal or northern fur seal, on all three pinniped species the $\delta^{15}N$ analyses indicated large differences in diet among individuals but no trends over time.

### Scat Collections

As noted above, collections of scat samples increased dramatically in the early 1990s. By spring 2002, thousands of scat samples were collected range wide and analyzed for seasonal and temporal shifts in prey parts, hormone and contaminant residues, and genetic composition. Based on about 400 scat samples, Merrick et al. (1997) concluded that the "high correlation between area-specific diet diversity and population changes supports the hypothesis that diet is linked with the Steller sea lion population decline in Alaska." This study forms the basis for the "diet diversity" hypothesis (see Chapter 6).

## CHALLENGES/LOGISTICS OF WORKING WITH SEA LIONS

Because of their seasonal, behavioral, and spatial distributions, this species is particularly difficult to study with a hands-on approach. During the nonbreeding season, the animals congregate at haulout locations that are distinct from rookery sites, although a limited number of animals may remain at a rookery location through the winter. There are many more haulout sites than rookery sites throughout their range and the animals disperse during the late summer from the rookery sites to the haulout sites. However, if adult females are not going to pup, they may remain at sea and not return to rookery sites. This attendance behavior is a critical limitation in monitoring pup production rates and female condition. Essentially, only those females healthy enough to produce pups arrive at the rookeries.

1. *Pups.* For studies that focus on newborn pups, these young animals are captured on their natal beaches up to about 3 to 4 weeks of age. After that, they are too fast, strong, wary, and easily able to move into the water to be captured in large numbers. Therefore, hands-on studies of newborn pups (branding, physiology, nursing behavior, etc.) must take place from mid-June to early/mid-July. By working within this time limitation, hundred of pups have been captured and handled on their natal beaches with large field teams.

2. *Juveniles.* Only 26 individuals less than 2 years of age were captured

and tagged from 1990 to 1996 (Alaska Department of Fish & Game and National Marine Fisheries Service, 2001). This is because until the very late 1990s it was difficult to capture juveniles (postpup, prebreeding) on the winter haulouts given the wariness of the animals and the difficulty of stalking and darting them in poor weather conditions. However, beginning in 1998, the Alaska Department of Fish & Game refined an underwater technique that allowed capture of as many as five young animals per day. There are now routine year-round expeditions by both the Alaska Department of Fish & Game and NMFS that capture young animals up to 2 to 3 years of age. As of the spring 2002, over 309 animals had been captured by the two agencies using this method (Thomas Gelatt, Alaska Department of Fish & Game, Anchorage, personal communication, 2001). This capture technique has been a major breakthrough in the study of young sea lions and has allowed initiation of several large programs of juvenile biology and behavior to be initiated.

3. *Adults*. Adult females can be captured on rookeries when they are moving back and forth from the sea while nursing their pups. Adult female Steller sea lions, like all otariids, make foraging trips to sea and then return to the beaches to nurse their pups. Among Steller sea lions, the first foraging trip by the female occurs about 7 days postpartum. The at-sea duration is typically less than 48 hours. Using this behavioral handle, it is possible to capture females when they are onshore with their pups so as to attach instruments and then have a reasonable chance of recapturing the females on a return trip. The caveat to such an approach, however, is that the sample size is low for the effort required. This method requires long stalking periods to find females on the rookeries that are good candidates for darting and large field teams consisting of anesthesiologists, animal handlers, boat drivers, and technical personnel. In addition, once a female is darted and the field team moves onto the rookery, the beach is heavily disturbed and other females typically leave the site and move into the water, leaving their pups onshore. The effort is such that no more than 5 or 6 females have been captured and handled in a single breeding season. This is not a method that is easily scaled up to large datasets and is a serious limitation for studies that focus on adult breeding females.

Working with adult females on their winter haulouts to gain data on condition, pregnancy status, or behavior during the nonbreeding season is extremely difficult because the animals are wary during this time, they have no pups waiting for them on the beach, and the winter weather makes procedures such as darting much more difficult. Consequently, no adult females have been captured during the winter among the eastern stock and only five have been captured from the Gulf and the Aleutians (Alaska Department of Fish & Game and National Marine Fisheries Service, 2001). Finally, fewer than six adult males have ever been captured at any time of the year because of their aggressive behavior, large size, and less accessible locations on the beaches.

4. *Location and behavior.* An important aspect of sea lion behavior is that many sea lions haul out on rocky ledges or pinnacles on which it is impossible to land a small boat. Thus, while sea lions may be routinely visible in large numbers to tour boats and seem to be easy to study, biologists cannot capture them. This behavior is also critical to the susceptibility of the species to shooting. Sea lions on ledges, rookeries, or haulouts do not react strongly to nearby boats and though it may be difficult or impossible to land on the beach and capture live animals, it is very easy to shoot at them from boats or a nearby beach. This susceptibility to shooting has been noted before as a particular issue with this species (Thorsteinson and Lensink, 1962). However, on a positive note, this behavior may favor new techniques that use darting methods to mark or take small biopsy samples from animals that otherwise could not be handled.

5. *Permit issues.* A final caveat is that the endangered status of Steller sea lions and increased legal, political, and economic ramifications of research associated directly with the species have made the Marine Mammal Protection Act permitting process cumbersome. Many hands-on research projects approved during the spring 2001 in the multimillion dollar Steller Sea Lion Research Initiative had still not been approved for permitting as of spring 2002.

6. *Demographic and temporal limitations.* Collecting biological information about the species (e.g., mass at age for adult females) is difficult because of the biological, handling, and legal issues addressed above. Thus, basic information that is necessary to define current conditions of the animals or to define lifetables must be collected indirectly, for example, with branding and resight studies with pups. Therefore, despite over $80 million allocated to collect such data, even an ideal experiment designed to define the survivorship of pups would take at least 7 to 8 years until the animals matured and survived to reproductive age. The basic biology of the species cannot be altered by legislation, money, or research. Any experimental or adaptive management method proposed must consider the time necessary to see the results. There must be a commitment to a multiyear approach because of the basic life history of the species.

# 5

# Fisheries

## GROUNDFISH FISHERIES

### Introduction

Fisheries conducted off Alaska's extensive 7,500-mile coastline can be divided into three very large regions—Gulf of Alaska, Aleutian Islands, and eastern Bering Sea (see Figure 5.1). The Gulf of Alaska is subdivided into eastern, central, and western areas. These areas are further subdivided into smaller federal reporting areas and even finer state statistical areas. A region of international waters exists in the central Bering Sea, beyond 200 nautical miles (nm) from the U.S. and Russian coasts, which is often termed the "Donut Hole." These designations differ from those used to describe the distribution of Steller sea lions.

Groundfish fisheries are of particular interest with respect to Steller sea lions. These fisheries are the main focus of this chapter because groundfish species comprise a large fraction of the sea lion diet (see Chapter 4). Groundfish include many fish species that are primarily distributed on or near the seafloor. The primary Alaskan groundfish species of commercial interest include walleye pollock (*Theragra chalcogramma*); Pacific cod (*Gadus macrocephalus*); Atka mackerel (*Pleurogrammus monopterygius*); sablefish (*Anoplopoma fimbria*); rockfish, especially Pacific ocean perch (*Sebastes alutus*) and several other rockfish species; and yellowfin sole (*Pleuronectes aspera*), and other flatfishes, including Pacific halibut (*Hippoglossus stenolepis*), which are not considered groundfish in the reporting of catch

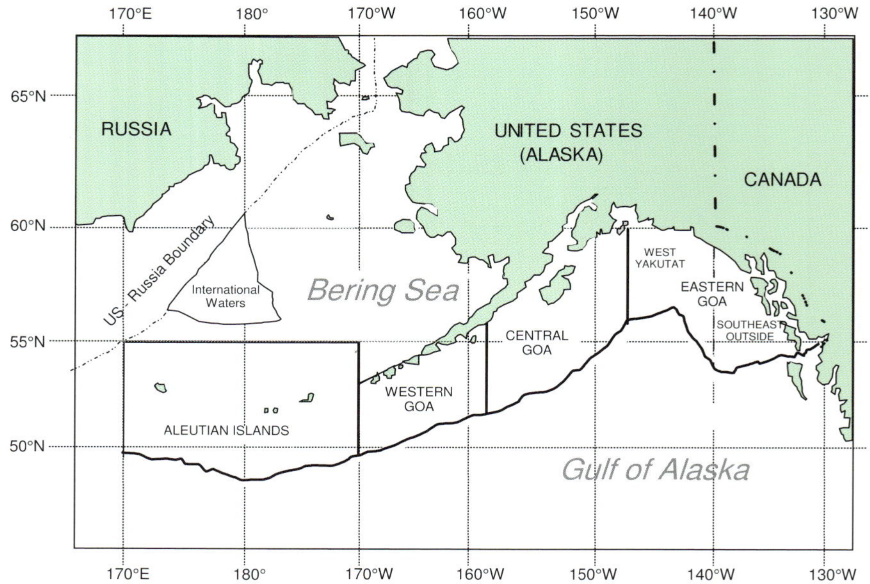


FIGURE 5.1 This map shows the principal federal regulatory areas for groundfish fisheries off the coast of Alaska extending to 200 nm offshore (solid wavy lines). The region in the central Bering Sea labeled "International Waters" is commonly referred to as the Donut Hole.
SOURCE: 50 *CFR* 679, Figure 14. Available from *www.fakr.noaa.gov/rr/figures.htm*, Sablefish Regulatory Areas and Districts.

statistics. Currently, pollock, Atka mackerel, and flatfish (except Greenland halibut, *Reinhardtius hippoglossoides*, which are caught mostly by longline) are harvested nearly entirely by trawl gear. Depending on the species, rockfish are taken largely by trawls, with smaller amounts taken on longline and jig gear. Pacific cod are harvested mostly by trawls and to a lesser extent by longlines, pots, and jig gear. Pacific halibut and sablefish are taken by longlines. Species abundance and the magnitude of the fisheries vary spatially. For instance, the Bering Sea accounts for more than 90% of Alaskan pollock harvests, whereas Atka mackerel are taken almost exclusively in the Aleutian Islands. For Pacific cod, the largest fishery takes place in the Bering Sea, although significant fisheries occur in the Gulf of Alaska and Aleutian Islands as well.

Given the diversity and complexity of commercial fisheries in the vast area covered by the Gulf of Alaska, Aleutian Islands, and Bering Sea, it is not possible to completely describe each fishery in this report. Instead, the focus is on particular aspects of fisheries that may be most relevant to considerations of the decline in sea lions. In the rest of this chapter, the strategy is first to review the history of the groundfish fishery with respect to Steller sea lion trends and, second, to describe the status of groundfish stocks and exploitation rates used to prescribe catch quotas under federal fishery management plans. Third, the spatial distribution of fisheries is considered, particularly with respect to the localized depletion hypothesis. Then, a brief overview is provided of some fisheries managed by the state of Alaska. Finally, management measures aimed at reducing potential competition between fisheries and Steller sea lions are summarized.

## Fishery History with Reference to Steller Sea Lion Trends

Alverson et al. (1964), Alverson (1992), Kruse et al. (2000), Rigby et al. (1995), and the National Marine Fisheries Service (2002) reviewed the history of groundfish fishing in Alaska. A brief overview is provided here. Although the focus of this chapter is the groundfish fishery, it is important to note that other species were very important in the early development of commercial fisheries in Alaska. In the Gulf of Alaska, Russian, Japanese, and domestic shrimp trawl fisheries developed in the early 1960s. By 1972, this fishery was fully Americanized. Domestic landings peaked in the mid-1970s, and shrimp stocks collapsed in the early 1980s. A Japanese trawl fishery developed in the Bering Sea in the later 1950s, peaked in the early 1960s, and ended in the later 1970s. Attempts to develop a domestic shrimp fishery in the Bering Sea failed, owing to depletion of shrimp resources. The histories of fisheries for salmon, crab, and herring are reviewed later in this chapter.

### Gulf of Alaska

Groundfish fisheries have been conducted off the Alaskan coast for nearly 150 years. In the 1860s a domestic schooner-based dory fishery using hand lines began for Pacific cod at various locations in the Gulf of Alaska (Bean, 1887). In the 1880s a longline fishery, conducted by U.S. fishermen, was developed for Pacific halibut and then sablefish in the inside waters of southeastern Alaska (Rigby et al., 1995). The sablefish fishery remained relatively unimportant until the mid-1930s, but halibut were depleted on some fishing grounds by 1916, leading to conservation concerns. In the 1960s Japanese longline fisheries developed for Pacific halibut, cod, and sablefish. Effort increased in the 1970s, and Korean vessels also participated in longline fisheries in the early 1970s.

Russian and, to a lesser extent, Japanese trawl vessels heavily exploited Pacific ocean perch (a slope rockfish species) in the early 1960s and 1970s, with a peak catch of 350,000 metric tons (mt) in 1965 (see Figure 5.2). This heavy exploitation occurred during a period when Steller sea lion populations were considered healthy and Pacific ocean perch rarely appeared in the sea lion diet. After a period of pulse fishing, Pacific ocean perch were overfished in the Gulf of Alaska, and Russian fleets moved south to exploit rockfish stocks off Oregon and Washington. High trawl bycatch, accounting for one-third of the total domestic longline halibut harvest in 1965, caused concern among domestic fishermen. Groundfish landings grew steadily through the 1970s and into the 1980s.

With the passage of the Magnuson Fishery Conservation and Management Act of 1976, the U.S. declared management authority over the Exclusive Economic Zone (EEZ) to 200 nm offshore, and the total allowable level of foreign fishing was limited to that portion of the optimum yield that was not expected to be harvested by domestic vessels. Americanization was promoted by a "Fish and Chips" amendment to the Magnuson Fishery Conservation and Management Act in 1980, which significantly raised foreign fishing fees and tied foreign fishing privileges to commitments to purchase products by the developing U.S. industry. The years 1978-1990 were a transitional period in which a number of joint foreign-

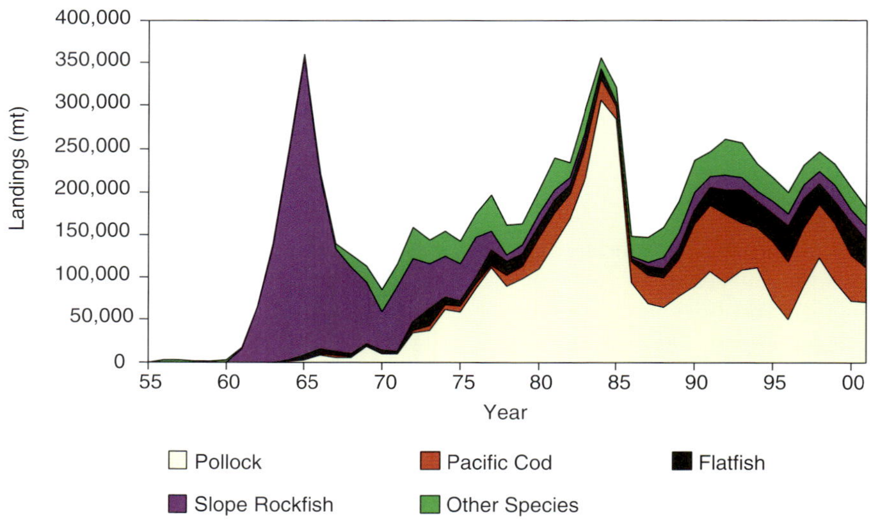


FIGURE 5.2 Groundfish landings (metric tons) from the Gulf of Alaska, 1956-2001. SOURCE: Graph prepared from data in Table 5, North Pacific Fishery Management Council (2002).

U.S. ventures were established whereby domestic fishermen operated catcher vessels that delivered to foreign processing motherships. A prime example was the pollock fishery in Shelikof Strait, a large body of water between Kodiak Island and the Alaska Peninsula in the central Gulf of Alaska (see Figure 2.1). After spawning aggregations were discovered there, a large joint-venture trawl fishery developed from 1981 to 1985 with primary emphasis on roe harvest and surimi production. Landings from this fishery alone peaked at 300,000 mt in 1984, when pollock accounted for 86% of all groundfish landings in the Gulf of Alaska. Pollock biomass declined significantly during 1982-1986, and landings were cut sharply starting in 1986. The pollock fishery became fully domestic in 1988, and all foreign and joint-venture fisheries were phased out of Alaska by 1991.

Shelikof Strait lies in the central Gulf of Alaska portion of the sea lion distribution. Although index counts of nonpup Steller sea lions declined 23% during 1976-1985 when the Shelikof Strait fishery was most intense, sea lions index counts for this region declined most sharply (63%) during 1985-1990, a period of reduced pollock harvest (landings averaged 79,500 mt annually during 1986-1990). Potential connections between this fishery and declines of Steller sea lions are discussed later in this chapter and also in the context of the "Human-Related Steller Sea Lion Kills" section in Chapter 6.

From 1990 to 2001, Steller sea lions continued to decline in the Gulf of Alaska, albeit at a reduced rate. During this same period, groundfish landings were relatively stable and averaged 229,000 mt, of which pollock averaged 91,000 mt, Pacific cod averaged 64,000 mt, flatfish averaged 30,000 mt, and the remainder was comprised mainly of sablefish and rockfishes. Perhaps the most significant fishery trend during this period was a doubling of flatfish landings, largely due to increased global demand and new foreign purchasing arrangements for domestically caught groundfish.

Most groundfish fisheries off Alaska are managed by the National Marine Fisheries Service (NMFS) under federal fishery management plans adopted by the North Pacific Fishery Management Council (NPFMC). For most groundfish fisheries in territorial waters (0 to 3 nm), state regulations parallel federal management actions. In some other instances, Alaska's Department of Fish & Game establishes separate catch quotas and other regulations for groundfish fisheries in state waters under management plans adopted by the Alaska Board of Fisheries. The state has lead management authority over lingcod (*Ophiodon elongatus*), black rockfish (*Sebastes melanops*), and blue rockfish (*S. mystinus*) throughout all waters of the territorial sea and the EEZ. Although these state-managed groundfish fisheries are important locally, they account for a small fraction of the total groundfish landings. In 2000 the largest state-managed fisheries were conducted for Pacific cod, and they accounted for 5% (12,265 mt) of the total (244,193 mt) cod landings off Alaska (Kristin Mabry, Alaska Department of Fish & Game, Division of Commercial Fisheries, Juneau, personal communication, 2001).

### Aleutian Islands

As in the Gulf of Alaska, the foreign trawl fishery depleted stocks of Pacific ocean perch in the Aleutian Islands in the 1960s and 1970s (see Figure 5.3). Landings of this species peaked at 115,000 mt in 1965. In the late 1970s a relatively small fishery developed for pollock, primarily in the eastern Aleutian Islands (Fritz, 1993). During 1980-1988 the fishery averaged 54,000 mt, peaked at 156,000 mt in 1989, and declined steadily from 74,000 mt in 1990 to 24,000 mt in 1998. In the late 1990s this fishery was increasingly constrained by regulations to avoid possible adverse effects on Steller sea lions. As a precautionary measure, the pollock fishery in the Aleutian Islands has been halted since the 1998 season.

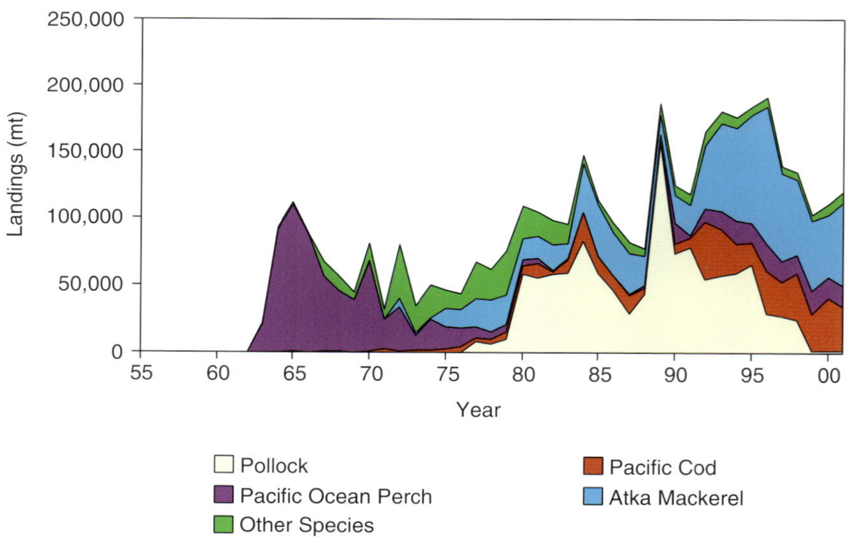


FIGURE 5.3 Groundfish landings from the Aleutian Islands, 1962-2001.
SOURCE: Graph prepared from data in Table 3, North Pacific Fishery Management Council (1999).

A foreign fishery developed for Atka mackerel in the 1970s with mean annual landings of 13,000 mt during 1972-1979. Joint-venture fisheries dominated the 1980s with average landings of 24,000 mt, but in 1990 U.S. domestic fishermen took over. Atka mackerel became the most important groundfish fishery in the Aleutian Islands in the 1990s as landings increased from 47,000 mt in 1992 to 103,000 mt in 1996 and averaged 55,000 mt during 1997-2000 (see Figure 5.3). The fishery is conducted in scattered locations, but most catches are taken from the central Aleutian Islands (Fritz, 1993).

In the 1960s, Japanese longline fisheries developed for Pacific halibut, cod, and sablefish. Effort increased in the 1970s, and Korean vessels also participated in longline fisheries in the early 1970s.

It is difficult to closely match periods of fishery development with periods of sharp declines of Steller sea lions in the Aleutian Islands. Declines of Steller sea lions appeared to begin in the eastern Aleutian Islands where nonpup index counts declined 61% during 1977-1985. Indeed, a significant pollock fishery first developed there during 1980-1985. However, pollock landings peaked in 1989 and continued at 1980-1985 levels through 1995, but sea lion index counts for this region increased 55% during 1989-1996. Likewise, sea lion counts in the central Aleutian Islands declined the sharpest (67%) during 1985-1989, and the Atka mackerel fishery has occurred in this area since the late 1970s (Fritz, 1993), but landings did not increase markedly until the 1990s.

During 1990-2001, groundfish landings in the Aleutian Islands averaged 145,000 mt per year. Average annual landings were dominated by Atka mackerel (58,000 mt), pollock (39,000 mt), cod (27,000 mt), and Pacific ocean perch (14,000 mt).

## Bering Sea

Japanese vessels conducted exploratory groundfish fisheries in the Bering Sea in 1929 and 1931. A fishery was conducted in 1933-1937; yellowfin sole was the primary target species. After a hiatus during World War II, the Japanese resumed a yellowfin sole fishery in 1954 (see Figure 5.4). A Russian fleet began fishing the eastern Bering Sea in 1958, and by 1960 combined total landings of yellowfin sole peaked at 500,000 mt annually. This fishery subsequently declined sharply by 1963, likely due to stock depletion.

In the early 1970s, foreign fleets of predominantly Japanese, Russian, and Korean vessels participated in Bering Sea groundfish trawl fisheries. About 450 foreign fishing vessels operated in the eastern Bering Sea in 1971. Reported foreign landings of pollock, flatfish, rockfish, and cod peaked at 2.1 million mt in 1972, of which 85% were pollock. Subse-

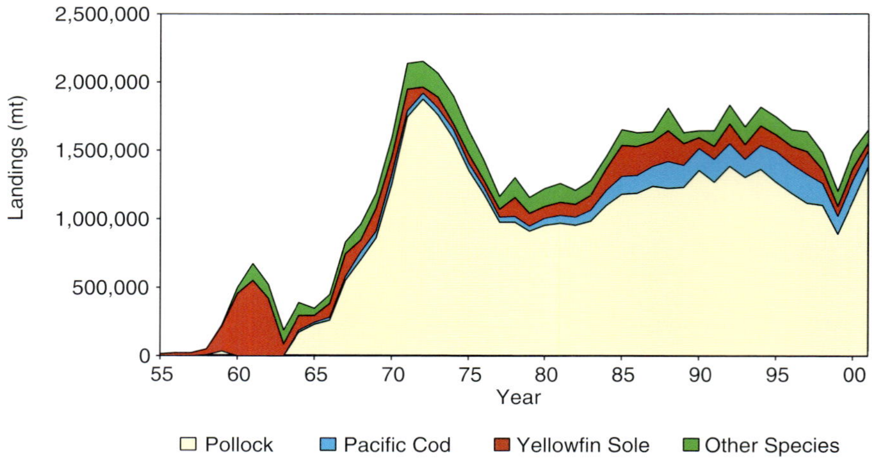


FIGURE 5.4 Groundfish landings from the Bering Sea, 1955-2001.
SOURCE: Graph prepared from data in Table 2, North Pacific Fishery Management Council (1999).

quently, total groundfish harvests declined, presumably due to reduced abundance.

The fishery recovered during the 1980s and landings have remained fairly constant since then. In the past decade (1990-2001), groundfish landings from the Bering Sea averaged 1.6 million mt, of which pollock accounted for 76% of the landings. Pacific cod accounted for 10%; yellowfin sole accounted for 7%; and the remainder included other flatfish, sablefish, and rockfish.

In the mid-1980s to early 1990s, a large pulse fishery for pollock developed in the Donut Hole in the central Bering Sea and a much smaller fishery developed near Bogoslof Island (north of the eastern Aleutian Islands) beyond the eastern Bering Sea shelf (see Figure 5.5). The Bogoslof fishery peaked at 377,000 mt in 1987, whereas the Donut Hole fishery peaked at 1.4 million mt in 1989. The latter was an intensive, essentially unregulated fishery with vessels from many foreign nations. Owing to severe depletion of pollock, most nations had stopped fishing by the time an international agreement was signed that closed the Donut Hole to fishing in 1993. Because of concerns about relationships between pollock near Bogoslof Island and those in the Aleutian Basin, the domestic Bogoslof Island fishery was closed in 1991.

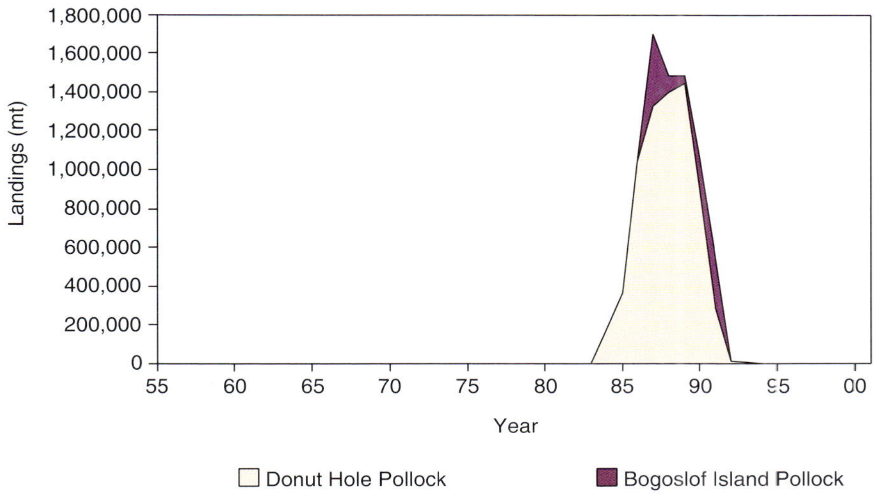


FIGURE 5.5 Pollock landings (metric tons) from the Bogoslof Island area and Donut Hole region in the central Bering Sea, 1955-2001.
SOURCE: Graph prepared from data in Table 1.12, Ianelli et al. (2001).

## Numbers of Fishing Vessels

The size and number of fishing vessels in addition to landings are important because they are the platforms by which humans interact with Steller sea lions and their prey. Northern Economics, Inc., and EDAW, Inc. (2002) reviewed fishing vessel participation in the federally managed groundfish fisheries of the Gulf of Alaska, Aleutian Islands, and Bering Sea for 1992-2000. During this period, the number of catcher-processor vessels (i.e., vessels capable of both catching and processing fish) declined 31% from 130 in 1992 to 90 in 2000 (see Figure 5.6a). Likewise, the number of groundfish catcher vessels (i.e., vessels that catch but do not process fish) declined 24% from 1,658 in 1992 to 1,261 in 2000. Most vessels that exited the groundfish fisheries were fixed-gear (nontrawl) vessels less than 60 feet in length. Significant numbers of vessels made small groundfish landings but are not included in these statistics. For instance, during 1992-2000, of the 4,403 vessels that landed at least some groundfish, 2,376 landed at least 70 tons of groundfish per year. Total landings from the remaining 2,037 vessels accounted for just 0.1% of the total volume of groundfish landings. Most of these vessels targeted other species but delivered some groundfish as bycatch.

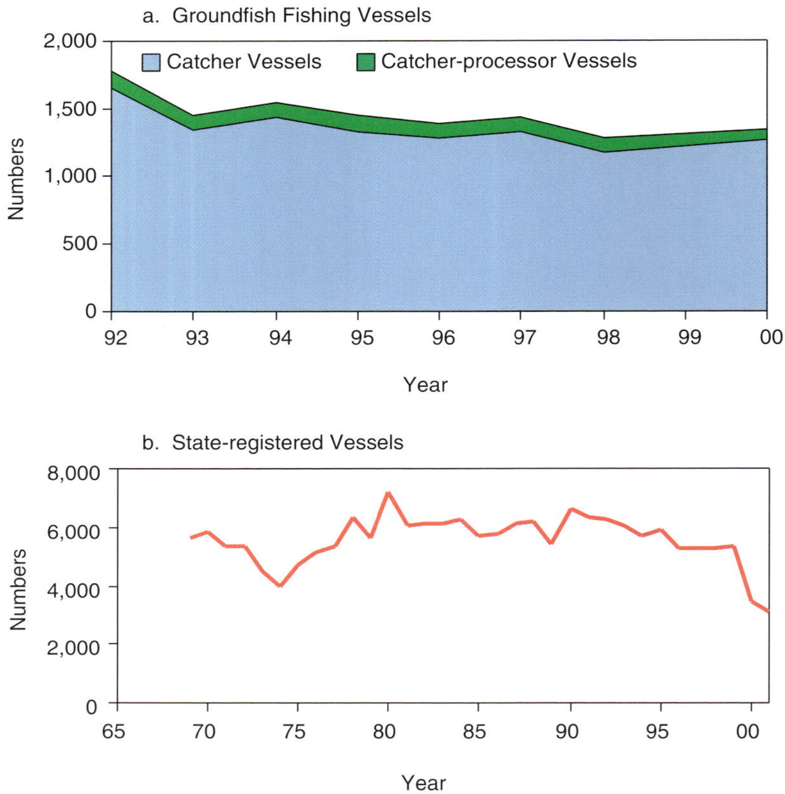


FIGURE 5.6 Number of vessels participating in (a) federal groundfish and (b) state fisheries in Alaska. (a) Catcher vessels and catcher-processor vessels that made significant landings (>70 tons annually per vessel) in the groundfish fisheries of the Gulf of Alaska, Aleutian Islands, and Bering Sea, 1992-2000 and (b) fishing vessels (all fisheries) registered with the state of Alaska in the region occupied by the western stock of Steller sea lions, which excludes southeastern Alaska and regions north of Bristol Bay, 1969-2001. Vessels must register with the state to fish for salmon, herring, crabs and other invertebrates, and groundfish (state waters only). Data for 2001 are preliminary. Numbers of state-registered vessels may be overestimated because vessel codes are unedited and may include errant identification numbers. Vessels participating in the halibut longline fishery are not included, and groundfish vessels fishing only in federal waters are not included unless they opt to register with the state.

SOURCES: Data for (a) from Northern Economics, Inc., and EDAW, Inc. (2002), Table 1.2-1, available at *www.fakr.noaa.gov/npfmc/Northern Economics/Northern Economics.htm.* Data for (b) provided by Susan Shirley, Alaska Department of Fish & Game, Juneau, personal communication, 2002.

Vessel size and type are other considerations in evaluating the effects of fishing vessels on the marine environment. Most catcher vessels can be divided into nine classes (Northern Economics, Inc., and EDAW, 2002). The participation of catcher vessels in groundfish fisheries off Alaska in 2000 was 590 (54.7%) nontrawl vessels of size 33 to 59 feet, 158 (14.7%) pot vessels greater than 59 feet, 72 (6.7%) longline vessels greater than 59 feet, 69 (6.4%) nontrawl vessels less than 33 feet, 46 (4.3%) trawl vessels less than 60 feet, 45 (4.2%) Bering Sea pollock trawlers of size 60 to 124 feet, 38 (3.5%) trawl vessels greater than 59 feet that are not eligible under the American Fisheries Act (AFA), 30 (2.8%) trawl vessels greater than 59 feet that are AFA eligible, and 30 (2.8%) trawl vessels greater than 124 feet.

Comparable time series of groundfish vessels have not been assembled prior to the 1990s owing to a lack of standardized data collection from foreign fishing fleets and lack of U.S. jurisdiction to 200 miles offshore prior to the Magnuson Fishery Conservation and Management Act of 1976. The state of Alaska has maintained data on fishing vessels since 1969. Vessels must register with the state to fish for salmon, herring, crabs, and other invertebrates in state and federal waters and groundfish (state waters only) or to enter state waters to deliver groundfish at an Alaskan port. Vessels participating solely in federally managed groundfish fisheries and delivering to offshore processors are not required to register with the state, and these vessels are underrepresented in the database. Participation was summarized for state-registered vessels for the region occupied by the western stock of Steller sea lions (see Figure 5.6b). Aside from a dip in the number of state-registered fishing vessels in the mid-1970s, participation has fluctuated around 6,000 vessels through the 1980s. As with the groundfish vessels, the number of state-registered fishing vessels declined 53% from 6,630 in 1991 to 3,108 in 2001. Counts of fishing vessels do not appear to reveal any striking patterns concurrent with Steller sea lion declines; however, the committee did not analyze trends at finer spatial scales or for particular fisheries.

## Groundfish Stock Status and Harvest Rates

Harvest rates have varied over time with changes in policy, improvements in stock assessments, and increased understanding of fish stock productivity (DiCosimo, 1998). For groundfish fisheries under the purview of the NPFMC, acceptable biological catch (ABC) and overfishing levels are prescribed under a six-tier system depending on the level of information available about stock status. Many fish stocks in the Gulf of Alaska, Aleutian Islands, and Bering Sea fall into tier 3 in which the ABC is set based on $F_{40\%}$, the theoretical rate of fishing mortality that reduces spawning stock biomass per recruit to 40% of the pristine (estimated unfished) level. The theoretical steady-state spawning stock biomass resulting from an $F_{40\%}$ harvest strategy is termed $B_{40\%}$. For modeled fish populations with mortality, growth, and recruitment parameters typical of groundfish such as pollock and cod, an $F_{35\%}$ strategy provides for high yields with low probability of reducing the fish stocks to dangerously low levels (Clark, 1991; Mace, 1994), and $F_{40\%}$ was chosen to be more conservative by providing a buffer against uncertainty and management error. The actual fraction of fish biomass removed under an $F_{40\%}$ harvest rate depends on life history characteristics that vary by species and stock as well as the variability in year-class strength and the fraction of immature fish not included in the spawning biomass calculations.

After considering estimates of ABC and other biological and economic information, the NPFMC sets a total allowable catch (TAC) to regulate commercial harvests for the year. For the Bering Sea the sum of all TACs must be no greater than 2 million mt; if this cap would be exceeded, then TACs for individual species (often pollock) are reduced so that the aggregate TAC stays within the prescribed limit of 2 million mt. Within the season, when the TAC for a particular species is reached (based on removals by the directed fishery and bycatch from other fisheries), that species becomes prohibited. Moreover, if continued fishing for other species creates a risk of overfishing (from bycatch), those other directed fisheries also could be suspended for the year.

Witherell et al. (2000) provide an example of the catch specification process for pollock in the eastern Bering Sea. Based on stock assessments conducted in 1999, projected biomass of age 3+ pollock (i.e., weight of all pollock age 3 and older; younger fish are not fully vulnerable to survey gear and cannot be accurately assessed) was 7.7 million mt. Although ABC was calculated to be 1.2 million mt using the $F_{40\%}$ strategy, fishery scientists recommended a more conservative harvest strategy owing to uncertainties in recruitment and other information. Thus, the NPFMC lowered the ABC to 1,139,000 mt and set TAC equal to ABC. In the Aleutian Islands a similar process was undertaken, although less information about that stock prevents use of the $F_{40\%}$ strategy. Instead, harvest rates are taken to be 75% of the natural mortality rate (M), resulting in an ABC of 23,800 mt for 2000. However, rather than setting TAC equal to ABC, TAC was set equal to zero, owing to low stock abundance and concern for Steller sea lions. For similar reasons, no fishing (TAC = 0) was allowed in the Bogoslof area.

The $F_{40\%}$ harvest strategy has been a source of some confusion, regarding its effect on the fish biomass and prey availability to Steller sea lions. It is not a fishing rate that depletes the entire fish stock to 40% of the level that would occur in the absence of fishing. Rather, it is a rate that, in theory, would reduce spawning stock biomass per recruit to 40% of unfished levels on average. In the case of the latest pollock stock assessment for the eastern Bering Sea (Ianelli et al., 2001), spawning biomass (i.e., the metric against which the $F_{40\%}$ rate is applied) was projected to be 3 million mt at the time of spawning (spring) in 2002. On the other hand, age 3+ biomass was estimated to be 9.8 million mt at the start of 2002. The 6.8-million-mt difference between the two biomass estimates is immature fish; only a small fraction of age 2 pollock are mature, and 100% maturity is not reached until age 10 (Ianelli et al., 2001). The biomasses of age 1 and 2 pollock are not included in either estimate because of gear selectivity. As sea lions consume both juvenile and adult pollock (Merrick and Calkins, 1996), an $F_{40\%}$ strategy does not deplete the entire pollock stock or even the portion consumed by sea lions to 40% of unfished levels. In fact, in the case of eastern Bering Sea pollock, the projected spawning biomass for 2002 is well above estimates of $B_{40\%}$ at 2.6 million mt and $B_{msy}$ (biomass corresponding to maximum sustainable yield) at 2.1 million mt.

Historical exploitation rates applied to fish stocks can be approximated from estimates of exploitable fish biomass and catch. For most Alaskan groundfish stocks, historical changes in fish biomass are reconstructed using age-structured analytical models, such as the Stock Synthesis Analysis (Methot, 2000), whereby historical data from assessment surveys, catch sampling programs, and landings records are assimilated with life history statistics in a model estimation framework. Time series of biomass and historical catches were collected from the recent stock assessment conducted in 2001 (North Pacific Fishery Management Council, 1999, 2002). For expediency the proportionate exploitation rate (i.e., the fraction of the exploitable biomass removed by the fishery) was estimated by simply dividing the catch into the annual exploitable biomass estimate. Only pollock, cod, and Atka mackerel were considered here because of their apparent importance to the Steller sea lion diet.

In the Gulf of Alaska, estimated biomass of pollock of age 2 and greater increased from 431,000 mt in 1961 to 4.1 million mt in 1981 and declined in the 1990s to levels typical of the 1960s and early 1970s (see Figure 5.7a). Dynamics of this stock are largely attributed to a series of three strong year classes in the early 1970s followed by five consecutive strong year classes in the mid- to late 1970s (Dorn et al., 2001). Since then the frequency of strong year classes has diminished. Likewise, four strong year classes during 1976-1980 (Thompson et al., 2001) led to increases in cod biomass during the late 1970s and early 1980s, although it has declined since 1990 due to below-average year classes (see Figure 5.7b). New survey data suggest that juvenile abundance is near record-high levels owing to one or two strong year classes, and the expectation is that the stock will increase over the next few years as these fish grow and mature (Dorn et al., 2001). Exploitation rates for both pollock and cod fluctuated between 5 and 15% during the 1980s and 1990s (see Figures 5.7c and d).

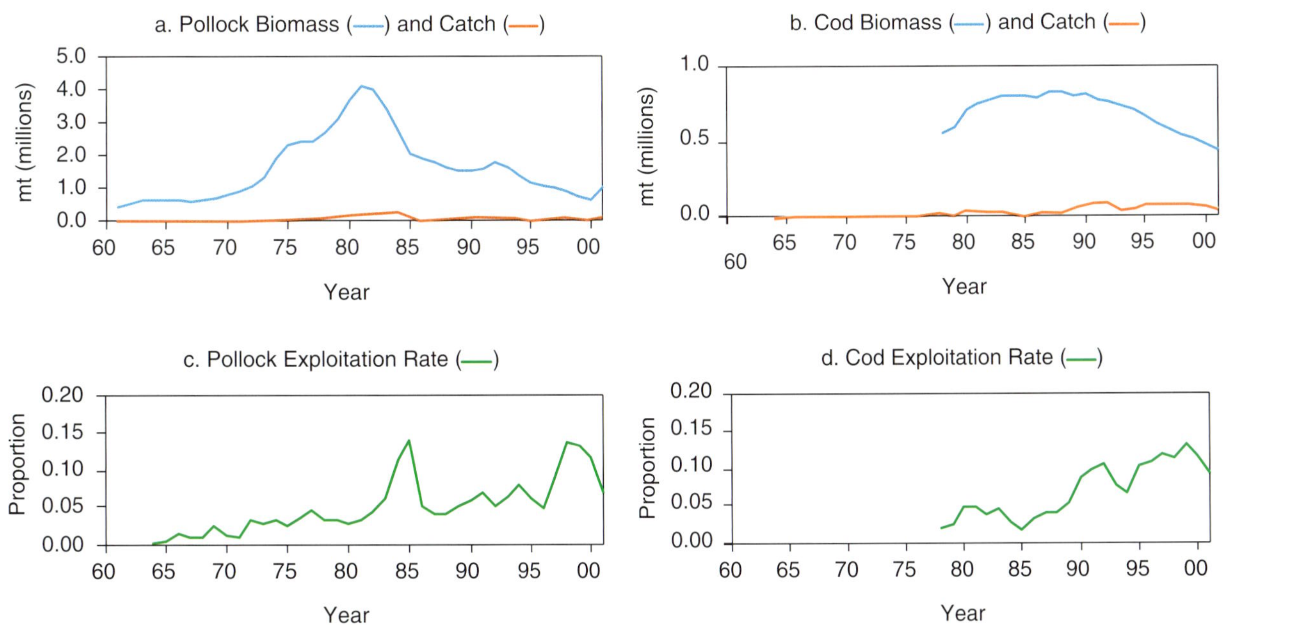


FIGURE 5.7 Gulf of Alaska groundfish exploitation, 1961-2001. (a) Estimated biomass of age 2+ walleye pollock and commercial catch, and (b) estimated biomass of age 3+ Pacific cod and commercial catch. Catch and biomass estimates are based on stock assessments conducted during 2001. Exploitation rate was approximated as estimated biomass divided by reported catch for (c) pollock and (d) cod.

SOURCES: Biomass data for pollock from Table 1.16, Dorn et al. (2001); data also available NMFS Alaska Fisheries Science Center website *www.refm.noaa.gov/stocks/specs/Data%20Tables.htm*. Catch data from Table 5, North Pacific Fishery Management Council (2002). Cod data from Table 2.24, Thompson et al., (2001); also available at the NMFS Alaska Fisheries Science Center website (see above).

In the Aleutian Islands, biomass estimates of Atka mackerel are highly variable because this species is patchily distributed on rough, rocky bottoms that are difficult to sample with trawl survey gear (Lowe et al., 2001). The species occurs primarily in the Aleutian Islands, but the 1994 and 1997 stock assessments located a significant biomass at a single survey station in the southern Bering Sea. Two strong year classes in 1977 and 1988 were largely responsible for an increase in biomass in the late 1970s and again in the early 1990s (see Figure 5.8a). During 1977-2001, the exploitation rate averaged 6.5%, with a peak of 14.2% in 1996 (see Figure 5.8b).

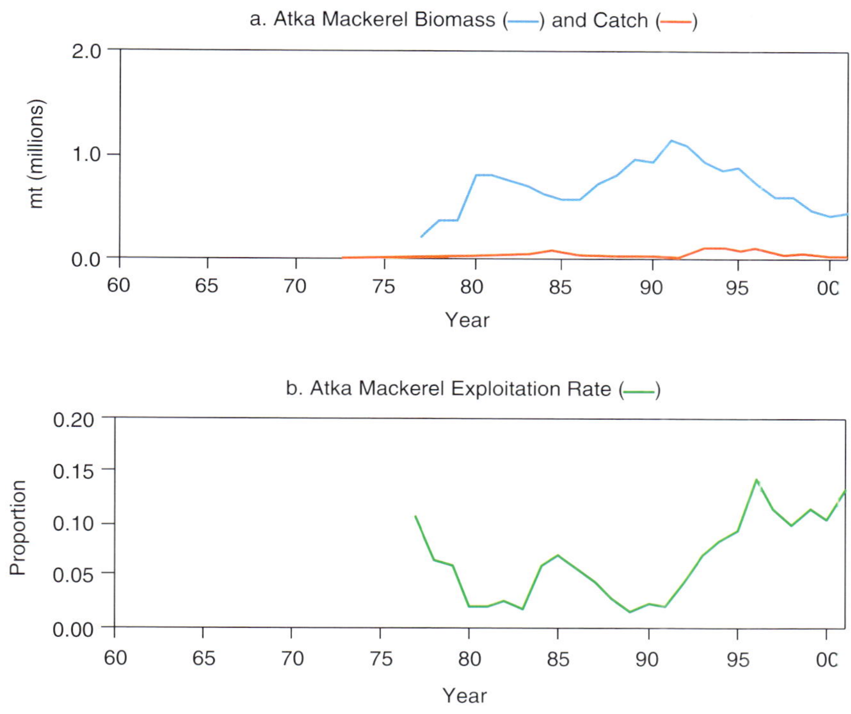


FIGURE 5.8 Aleutian Islands and Bering Sea Atka mackerel exploitation, 1977-2001. The top graph (a) shows estimated age 3+ biomass and commercial catch. Catch and biomass estimates are based on stock assessments conducted during 2001. Exploitation rate was approximated as estimated biomass divided by reported catch.
SOURCES: Biomass data from Table 13.9, Lowe et al. (2001); data also available at the NMFS Alaska Fisheries Science Center website *www.refm.noaa.gov/stocks/specs/Data%20Tables.htm*. Catch data from Table 3, North Pacific Fishery Management Council (1999).

The eastern Bering Sea supports by far the largest pollock fishery in Alaska, so trends in biomass and exploitation rate are of particular interest with respect to sea lions. Estimated biomass of pollock averaged 3.9 million mt during 1964-1980, increased more than 2.5-fold during the early 1980s, and averaged 10.4 million mt during 1983-2001 (see Figure 5.9a). Strong year classes in 1978, 1982, 1984, 1989, 1992, and 1996 are associated with subsequent periods of high abundance (Ianelli et al., 2001). Exploitation rates averaged 12.3% during 1980-2002 (see Figure 5.9c). Estimates of exploitation rate are questionable for the early years of this fishery because of uncertainty in the biomass estimates. For example, the exploitation rate in 1974 was approximately 50% based on the stock assessment conducted in 2001 and 94% based on the stock assessment conducted in 2000 (Ianelli et al., 2001). For the most recent 3 years, biomass estimates are also uncertain because young fish are only partially recruited to the fishery and are difficult to survey.

Estimated biomass of Pacific cod in the eastern Bering Sea and Aleutian Islands (note: most catch is taken in the Bering Sea) increased by an order of magnitude during the late 1970s to early 1980s and has been declining steadily since 1987, with a fairly steady increase in exploitation rate until 1997 (see Figures 5.9b and d). The biomass increase was associated with four strong year classes during 1977-1982, and the subsequent decline is associated with nine weak year classes during 1986-1998.

## Spatial Distribution of Harvests

Fishing effort is unevenly distributed off the coast of Alaska (Fritz, 1993; Fritz et al., 1998; National Research Council, 2002). The concentration of fishing effort in some areas has focused concern on the potential for fishing to cause localized depletion of prey species for Steller sea lions.

With the shift from foreign to domestic groundfish fisheries in the 1980s, the fraction of pollock, cod, and Atka mackerel landings in Steller sea lion critical habitat increased (see Figure 5.10). Critical habitat is defined by the 20-nm radii around sea lion rookeries and important haulouts, and three aquatic foraging areas in Shelikof Strait, Bogoslof Island area, and Sequam Pass. This shift in effort was confounded by increases in fish biomass in the 1980s, and it is difficult to evaluate potential effects on local prey densities from these data alone.

Fritz (1993) compiled maps of trawl locations from observer data collected aboard foreign, joint venture, and domestic fisheries for pollock in 1977-1992. It is difficult to accurately summarize the temporal and spatial variability in trawl locations in the eastern Bering Sea because of large variability by year and season. During 1977-1980, much of the observed hauls occurred along the continental shelf break, with some scattered

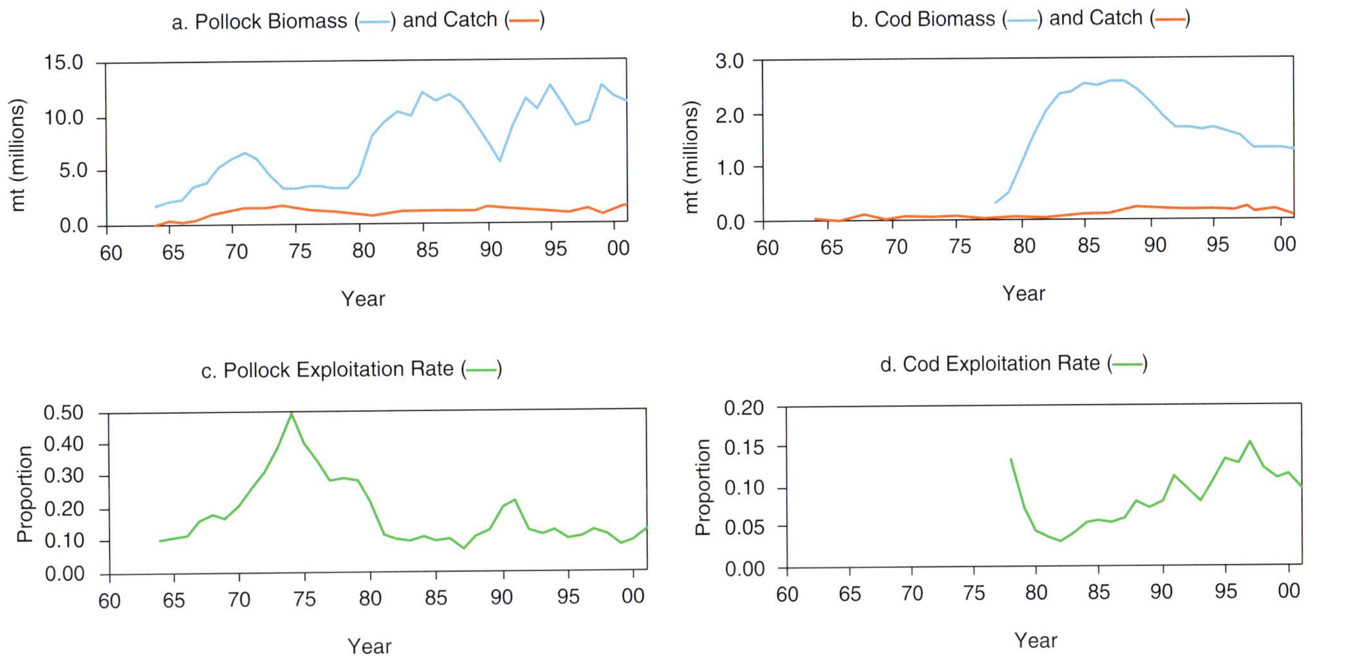


FIGURE 5.9 Bering Sea groundfish exploitation, 1964-2001. (a) Estimated biomass of age 3+ walleye pollock commercial catch, and (b) estimated biomass of age 3+ Pacific cod and commercial catch. Catch and biomass estimates are based on stock assessments conducted during 2001. Exploitation rate was approximated as estimated biomass divided by reported catch for (c) pollock and (d) Pacific cod.

SOURCES: Pollock biomass data from Ianelli et al. (2001). For cod, biomass data from Table 2.24, Thompson and Dorn (2001); data also available at the NMFS Alaska Fisheries Science Center website *www.refm.noaa.gov/stocks/specs/Data%20Tables.htm*. Pollock and cod catch data from Table 2, North Pacific Fishery Management Council (1999).

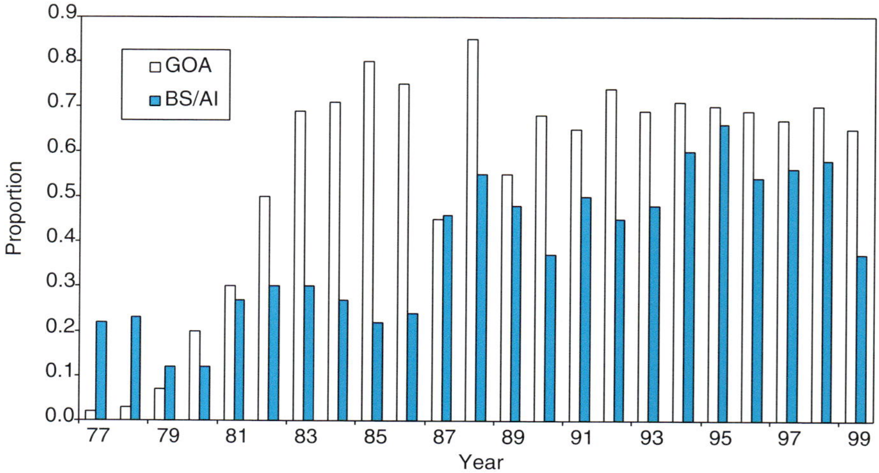


FIGURE 5.10 Proportion of pollock, Pacific cod, and Atka mackerel catches in Steller sea lion critical habitat for the Gulf of Alaska (GOA) and Bering Sea and Aleutian Islands (BS/AI) areas, during 1977-1999.
SOURCE: Data from National Marine Fisheries Service (2000).

effort on the continental shelf. During the early 1980s, effort expanded off the shelf in the Aleutian Basin from January to April, and effort expanded onto the continental shelf from May to December. From the mid-1980s to early 1990s, the geographic distribution of trawl locations was quite variable. For example, from January to April 1985 effort was concentrated mostly on the southeastern Bering Sea shelf in a relatively small area north of Unimak Island. From January to April 1986 nearly all effort occurred in the Aleutian Basin but shifted from May to August 1986 to the shelf north of 56° N with most effort on the northwestern portion of the shelf north of 58° N. After the closure of the Bogoslof Island fishery in 1992, most of the roe season (January to March) has been concentrated north and west of Unimak Island (Ianelli et al., 2001). Oceanographic factors, such as temperature, appear to influence the geographic distribution of both juvenile and adult pollock on the eastern Bering Sea shelf (Swartzman et al., 1994; Ianelli et al., 2001). In summary, survey and fishery data suggest a very high level of movement by pollock in the eastern Bering Sea.

In the Gulf of Alaska, observers documented several major shifts in the distribution of the pollock fishery during the late 1970s to early 1990s (Fritz, 1993). During 1977-1981, most pollock trawling occurred during

May to December along the continental shelf break. The distribution of observed trawl locations from September to December 1978 (see Figure 5.11a) typifies this pattern. In January to April 1982 an intense roe fishery developed in Shelikof Strait. In the same year, trawl effort continued and intensified along the shelf break during May through December. This new pattern continued until 1984, when the Shelikof Strait roe fishery became most intense (see Figure 5.11b), and fishing effort in May to December 1984 encompassed both the shelf break and portions of the outer continental shelf, particularly off Kodiak Island and the Alaska Peninsula (see Figure 5.11c). Both the winter roe fishery in Shelikof Strait and the non-roe fishery on the outer shelf and shelf break diminished substantially in intensity by 1986. This drop resulted from a combination of declining abundance and reduction in harvest rate (see Figure 5.7a, c). During 1990-1992, Shelikof Strait no longer dominated trawl locations during January to April to the same degree as during 1982-1986 (see Figure 5.11d), and during the rest of the year fishing effort was patchily distributed over the continental shelf in the central and western Gulf of Alaska with some concentrations in gullies off Kodiak Island (see Figure 5.11e).

The pollock roe fishery was intense for 5 years (1981-1985) in Shelikof Strait, and this area appears to be an important foraging area for Steller sea lions. It is uncertain whether these major shifts in fish removals adversely affected the number and density of pollock schools and the foraging success of Steller sea lions in the region. However, as described more fully under "Human-Related Steller Sea Lion Kills" in Chapter 6, thousands of sea lions were captured by this fishery (Perez and Loughlin, 1991) and many more were shot by fishermen to protect their catches and gear as nets were towed near the surface for delivery to motherships for processing. Summer bottom-trawl survey data from 1984 to 1996 show that there was no reduction in stations with high concentrations of pollock in the area (Shima et al., 2002). Interestingly, as the stock declined through the 1990s, the geographic distribution expanded, albeit at low densities, into deeper (200 to 300 m) waters than observed in the 1980s. This appears counter to expectations that pollock distributions would contract with reduced abundance as a density-dependent response consistent with MacCall's (1990) basin model. The spatial expansion of pollock into deeper waters when abundance declined may simply reflect an expansion of suitable habitat as ocean temperatures increased following the 1976-1977 regime shift.

In recent years a few studies have been conducted to examine the potential for localized depletion of sea lion prey by fishery removals. An ongoing study is attempting to determine whether commercial fishing adversely affects the distribution and abundance of pollock and other

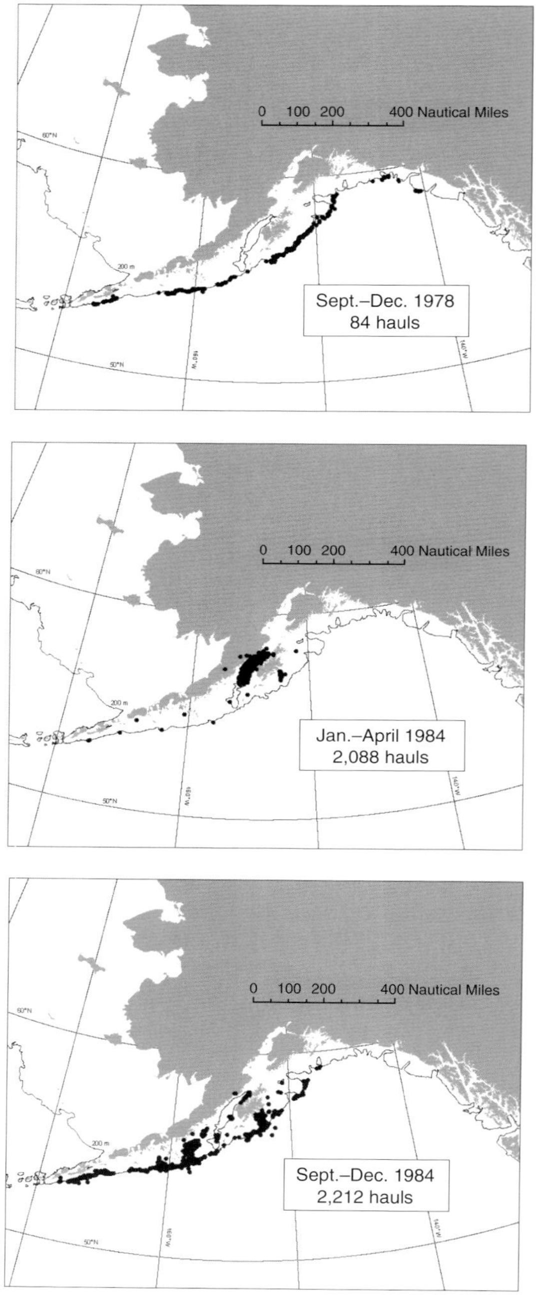


a

b

c

FIGURE 5.11 Continued on next page.

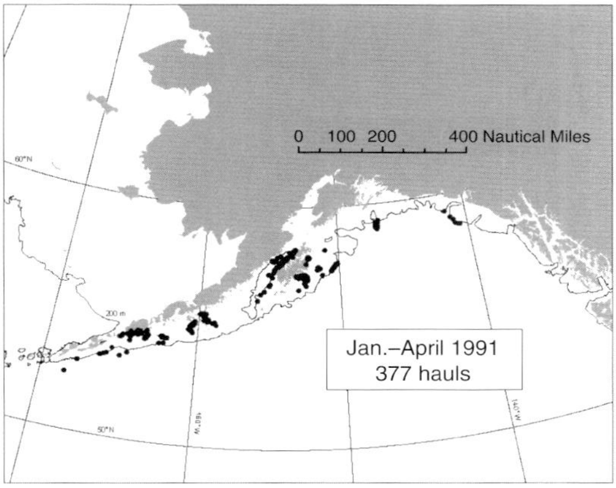


d

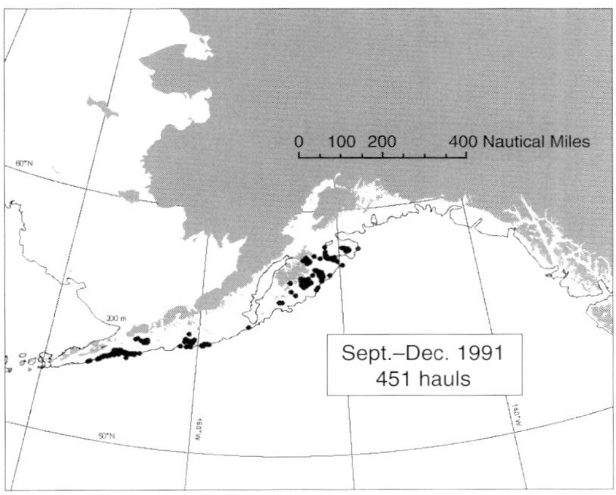


e

FIGURE 5.11 Locations of observed pollock trawl hauls in the foreign and joint-venture fishery in the Gulf of Alaska in (a) September-December 1978, (b) January-April 1984, (c) September-December 1984, and in the domestic fishery during (d) January-April 1991, and (e) September-December 1991.
SOURCE: From Fritz (1993); (a) p. 97, (b) p. 113, (c) p. 115, (d) p. 128, (e) p. 130.

forage fish over short spatiotemporal scales (Wilson et al., 2002). Two areas were studied off Kodiak, one open to trawling (Barnabas trough) and one closed (Chiniak trough). Hydroacoustic surveys were conducted before and during the fishery in 2001. No significant differences in the biomass of juvenile or adult pollock were detected in response to commercial fishing in which 27 vessels harvested 2,850 mt of pollock from Barnabas trough. Moreover, analysis of pollock school dynamics (e.g., shape, vertical distribution) indicated no significant differences between prefishing and fishing periods. In summary, there appears to be no significant link between this trawl fishery and the availability of pollock to Steller sea lions in a relatively small area on short timescales. However, these results are preliminary and reflect only one investigation into the localized depletion issue.

Other Steller sea lion prey species may be more susceptible to localized depletion. Evidence of seasonal, localized depletion of Atka mackerel by the trawl fishery in 1992-1995 has been reported by Fritz (1999). This fishery is concentrated in relatively few fishing areas (Fritz, 1993). In eight localized fisheries in the Aleutian Islands and Gulf of Alaska, Leslie regression analyses showed a sharp reduction in fishery catch per unit effort in some years and areas (Fritz, 1999). In areas with statistically significant declines, estimates of localized harvest rates were 55 to 91%. Fritz (2001) found significant depletion again at some sites in 1996 and 1997. In the Aleutian Islands, catch per unit effort returned to initial levels by the start of the fishery the following year, suggesting that the depletion was temporary. A more persistent depletion was observed from 1993 to 1994 at the Gulf of Alaska sites, but length-frequency distributions suggested that immigration of small fish contributed to the population. Preliminary results from a subsequent study indicate little movement of tagged adults in summer within 40 days of release, whereas a small proportion moved between areas within 64 days of release (Fritz et al., 2001). Because of concerns about localized depletion of Atka mackerel in areas important to foraging Steller sea lions, the NPFMC adopted regulations in June 1998 to disperse this fishery both temporally and spatially (Lowe et al., 2001). Now, the TAC is equally split into two fishing seasons—January 1 to April 15 (A season) and September 1 to November 1 (B season). A 4-year schedule was implemented to reduce the percentage of harvest in sea lion critical habitat in the central and western Aleutian Islands from 80% (1999) to 40% (2002). No similar reduction plan was adopted for the eastern subarea, but 20-nm trawl exclusion zones around Seguam and Agligadak rookeries, previously in effect for pollock trawling only, are now in effect year-round for the Atka mackerel fishery. Once this implementation schedule is completed, it would be prudent to conduct a follow-up to Fritz's (1999; Fritz et al., 2001) study to evaluate the efficacy of these management actions that were intended to eliminate apparent localized depletion. A similar study of localized depletion on walleye pollock in the Bering Sea is currently in progress, and results are not yet available (Brian Battaile, University of Alaska, Fairbanks, personal communication, June 27, 2002).

## Summary of Potential Effects of the Groundfish Fishery on Steller Sea Lions

Evidence linking the decline of the western stock of Steller sea lions to the development of commercial groundfish fisheries is equivocal. There is much uncertainty about the effects of fishing on fish prey on spatial and temporal scales associated with foraging sea lions due to a lack of directed research. With few exceptions, little is known about the effects of fishing on fish school dynamics and sea lion foraging behavior. By necessity, the committee focused on available information, which primarily includes historical regional trends of Steller sea lions, fish stock biomass, and fishery removals, and a few studies of fish spatial distributions and potential localized depletion by fisheries. From this review the following general conclusions are drawn.

There appears to be some evidence for overfishing of halibut in some areas of the Gulf of Alaska in the early 1900s: Pacific ocean perch in the Gulf of Alaska and Aleutian Islands in the 1960s and 1970s, yellowfin sole in the eastern Bering Sea in the 1960s, and walleye pollock in the central Bering Sea in the late 1980s. However, in each case the extent of overfishing cannot be quantified owing to the lack of assessment surveys on these fish stocks during those years. Depletion of pollock by a multinational fishery in the Donut Hole could have directly reduced prey availability to some Steller sea lions.

Although satellite telemetry data and at-sea observations (National Marine Fisheries Service, 2000) indicate that most sea lions stay on the continental shelf, some animals have been observed in the central Bering Sea. The diet of Steller sea lions in the Aleutian Islands is dominated by Atka mackerel, but depletion of pollock could further reduce diet diversity, which appears to be inversely correlated to the rate of decline (Merrick et al., 1997). Aside from pollock, other instances of overfishing occurred prior to the rapid decline of the sea lions in the 1980s and involved fish species that are less important components of the sea lion diet (see Chapter 4). For these earlier overfishing events to have played a significant role in sea lion declines, the effects must have been indirect and lagged over decades. Such potential ecosystem effects are evaluated in Chapter 6.

The intense Shelikof Strait pollock fishery in the early to mid-1980s may be implicated in declines in sea lions in the central Gulf of Alaska during 1976-1985. Undoubtedly, trawl bycatch and shooting of sea lions contributed to the decline in this region in these years, as discussed under "Human-Related Steller Sea Lion Kills" in Chapter 6. However, aspects of the pollock stock biomass and fishery history do not fit well with the decline of Steller sea lions in the central Gulf of Alaska in the context of a fishery competition hypothesis. For instance, pollock biomass increased during the mid-1970s to early 1980s (see Figure 5.7a), when the sea lion decline appears to have started (see Chapter 3). Counts of sea lions in this region declined most sharply during 1985-1990, after this fishery was substantially reduced, but there may have been a lagged effect of the fishery.

Pollock biomass declined from the early 1980s through the 1990s in concert with sea lion declines in this region. Yet pollock biomass in the late 1990s was similar to biomass levels in the 1960s to early 1970s when sea lion populations were larger. Biomass of another important gadid prey, Pacific cod, was high in the 1980s and began a decade of decline around 1990 (see Figure 5.7b). It is difficult to evaluate whether localized depletion occurred with these changes in fish biomass. Survey stations with high densities of pollock persisted during the pollock decline of the mid-1980s and 1990s, at least in summer when the assessments were conducted (Shima et al., 2002). Depletion may have occurred during winter fisheries but subsequent seasonal movements of fish prior to the summer survey may have masked this effect. However, preliminary results from an ongoing experimental study off Kodiak Island suggest that localized depletion may not occur within timescales of days to weeks during the fishing season (Wilson et al., 2002), at least under the conditions of that experiment.

The best-supported case for localized depletion has been made for Atka mackerel in the 1990s. Catches and exploitation rates increased in the 1990s. Also, Atka mackerel are vulnerable to localized depletion because of their schooling behavior and apparently limited movement. The sharpest decline in the central Aleutian Islands, where Atka mackerel are fished, occurred in 1985-1989, a period of low exploitation rates. Unfortunately, no information on localized depletion was developed during these years. In the 1990s the recovery of catch rates to prefishery levels by the start of the next fishing season suggests that within-season localized depletion is the concern, not overfishing.

In general, the fishery competition hypothesis is not supported by comparisons of trends in Steller sea lion populations and fish biomass in the Gulf of Alaska, Aleutian Islands, and eastern Bering Sea. The 1970s was a decade of high sea lion abundance despite relatively low groundfish abundance. The 1980s was a decade of high groundfish biomass notwithstanding sharp declines in Steller sea lions. Biomass of eastern Bering Sea pollock, the largest of all fisheries, declined to intermediate levels during 1987-1991 but rebounded by 1993 and has since remained at or near all-time record high levels of abundance. Likewise, Atka mackerel biomass increased in the 1980s and peaked in the early 1990s, a period of sharp declines in the sea lion population in the Aleutian Islands. Most of these shifts in groundfish biomass are related to changes in recruitment success, which appear to be largely driven by environmental conditions, rather than effects of fishing, in the range of observed fish biomass (e.g., Hollowed and Wooster, 1992, 1995).

An evaluation of groundfish harvest policy indicates that contemporary groundfish fisheries in the Gulf of Alaska, Aleutian Islands, and Bering Sea are generally managed conservatively based on stock assessments and reference points for target and overfishing rates. None of the groundfish stocks in the Alaska region are currently considered to be overfished (National Marine Fisheries Service, 2002). Errors in assessment are of concern, potentially leading to higher-than-intended harvest rates in some years, but even with this caveat, historical exploitation rates appear to be lower than overfishing rates with few exceptions over the past several decades. Accordingly, as discussed later in this chapter, sea lion protective measures have focused less on the overall rate of harvest and more on temporal and spatial apportionments of catches so as to mitigate the potential for localized effects by the groundfish fishery.

## STATE-MANAGED FISHERIES

### Overview

In addition to the groundfish fisheries, the state of Alaska manages other fisheries, including salmon, herring, crab, shrimp, scallops, and other invertebrates. Federal fishery management plans provide frameworks for state management authority for scallop and salmon fisheries statewide and for crabs in the Bering Sea and Aleutian Islands region. For other species the state manages fisheries within territorial and EEZ waters without the auspices of federal fishery management plans. Brief summaries are provided below of the historical and current herring, crab, and salmon fisheries managed by the state of Alaska following the more detailed review provided by Kruse et al. (2000).

### Herring Fisheries

Fisheries for herring are considered here because of their relative importance in the diet of Steller sea lions. Herring are one of the top 10 prey of sea lions, based on frequency of occurrence, although they generally occur in 10% or less of stomachs (Pitcher, 1981) or scats (National Marine Fisheries Service, 1999). Their contribution to the diet varies both seasonally and regionally. Thomas and Thorne (2001) observed aggregations of Steller sea lions feeding on herring schools at night in late winter in Prince William Sound. Fishery managers with the Alaska Department of Fish & Game commonly utilize the co-occurrence of sea lions and whales as indications that herring are approaching nearshore waters for spawning in spring (Kruse et al., 2000). The junk food hypothesis posits that sea lion nutrition was compromised by a regime shift in the late 1970s, which resulted in decreased abundance of herring and other fatty forage species and increased abundance of pollock and other groundfishes of lesser nutritional value (Alverson, 1992; Rosen and Trites, 2000).

Alaska's commercial herring industry began in 1878, and catch records are available dating back to 1915 (see Figure 5.12). Starting in the late 1800s, herring were rendered for oil at reduction plants. The fishery developed in southeastern Alaska, and this region dominated the catches in 1930, when it accounted for roughly 75,000 mt of the total 80,000 mt reported for Alaska. As herring stocks in southeastern Alaska declined,

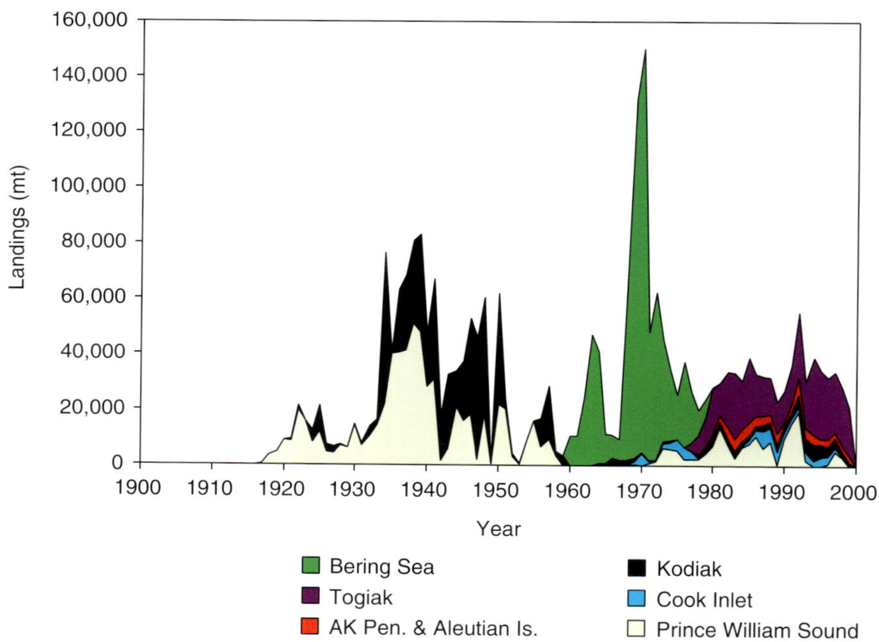


FIGURE 5.12 Historical herring catches from the Gulf of Alaska (excluding southeastern Alaska), Aleutian Islands, and Bering Sea. Catches include those from reduction fisheries prior to 1960, offshore foreign trawl fisheries in the Bering Sea during 1960-1980, and recent sac roe fisheries since 1970. AK Pen & Aleutian Is. includes catch data from Dutch Harbor (Aleutian Islands), and North and South Peninsula (Alaska Peninsula).

SOURCE: Data from Fritz Funk, Alaska Department of Fish & Game, Juneau, personal communication (2002).

the fishery shifted to Prince William Sound and Kodiak areas. Overall, the reduction fishery peaked at 142,000 mt in 1934. Catches generally declined but were highly variable through 1960, when they were discontinued. A large foreign fishery took place in the Bering Sea from 1960 to 1980. The Russians developed a trawl fishery in the late 1960s, and the Japanese developed trawl and gillnet fisheries in the early to mid-1960s. Catches peaked at over 140,000 mt in 1970. The contemporary fishery is conducted on prespawning fish to harvest their valuable roe for Asian markets. Harvests from the roe fisheries were relatively stable from 1980 to the mid-1990s (see Figure 5.12). In the late 1990s, catches declined sharply, with reductions in herring biomass, particularly in Prince William Sound, Cook Inlet, and Kodiak areas. Herring biomass and catches have remained relatively stable in southeastern Alaska and the eastern Bering Sea.

Herring are harvested in 25 distinct fisheries, almost all of which are tied to specific herring spawning locations. Contemporary fisheries are conducted by gillnet and purse seines. Fisheries are managed with catch quotas typically based on a harvest rate of 20% of estimated spawning biomass. Fishery thresholds, stock levels below which no fishing occurs, are commonly established at 25% of estimated unfished biomass. Fisheries have been closed in Prince William Sound since 1998 and lower Cook Inlet since 1999, because herring abundance has been at or below threshold.

Whether fisheries have depleted herring populations to the detriment of Steller sea lions remains an open question (see Chapter 6). The recent declines in herring abundance in Prince William Sound and Cook Inlet in the late 1990s occurred well after the declines in sea lions, but these declines are worrisome owing to the apparent importance of herring and squids in the diet of sea lions in Prince William Sound (Pitcher, 1981). Conversely, the large reduction fishery in the Gulf of Alaska peaked in 1934 and ended in 1960—well before the steep sea lion declines in the 1980s. Bering Sea herring stocks declined sharply in the late 1960s to mid-1970s associated with a large offshore trawl fishery and senescence of strong 1957-1958 year classes (Wespestad, 1991; Wespestad and Gunderson, 1991). Associated with recruitment of strong 1977-1978 year classes and termination of the trawl fishery, Bering Sea herring biomass recovered to average levels in the early to mid-1980s (Wespestad, 1991; Funk and Rowell, 1995), a period of decline of Steller sea lions in the Aleutian Islands. Since the late 1980s, herring biomass slowly declined (Funk and Rowell, 1995).

## Crab Fisheries

Crabs are not preyed on by Steller sea lions. However, as described in Chapter 6, sea lions are known to puncture inflated crab pot floats, prior to the introduction of solid-core Styrofoam "sea lion" buoys to prevent lost gear. In the 1960s and 1970s, anecdotal reports indicated that it was not uncommon for crab fishermen to shoot sea lions both in defense of their gear and for use as crab bait (Ed Opheim, former cod and salmon fisherman, Kodiak, personal communication, September, 2001). Because many of these fisheries are now closed due to reduced crab abundance, the historical development of Alaskan crab fisheries is of interest regarding a potential pulse of sea lion mortality associated with shootings by crab fishermen. Here, we only consider crab fisheries conducted in the range of the western stock of Steller sea lions. In southeastern Alaska, a portion of the region occupied by the eastern stock of Steller sea lions, there are small, but relatively stable, fisheries for red king, golden king, Tanner, and Dungeness crabs. Also, a small red king crab fishery occurs in Norton Sound.

The Japanese pioneered Alaskan crab fisheries with the tanglenet fishery in the 1930s. After a hiatus in the 1940s, the fishery resumed in the eastern Bering Sea in the early 1950s and the Russians entered the fishery in the late 1950s. The Japanese also caught crabs with pot gear in the late 1960s and early 1970s. These foreign crab fisheries continued until the early 1970s.

Domestic red king crab (*Paralithodes camtschaticus*) fisheries developed in Alaska in the mid-1950s to early 1960s (see Figure 5.13a). The Kodiak fishery peaked at 45,000 mt in 1965. Smaller fisheries were prosecuted in lower Cook Inlet and the Aleutian Islands. As red king crab fisheries declined through the 1970s in the Gulf of Alaska, a large fishery developed in Bristol Bay in the eastern Bering Sea. The Bristol Bay fishery peaked at 60,000 mt in 1980. This fishery plummeted in the early 1980s and was closed for a few years owing to depressed stock conditions. All red king crab fisheries have been closed since 1983, except for Bristol Bay and two small fisheries in southeastern Alaska and Norton Sound. Small (annual landings <5,000 mt) blue king crab (*Paralithodes platypus*) fisheries took place around the Pribilof Islands and St. Matthew Island in the eastern Bering Sea during the mid-1960s to late 1990s.

Japanese fishermen developed a blue king crab fishery off the Pribilof Islands in 1965, and U.S. domestic fishermen entered the fishery in 1973 (see Figure 5.13b). Domestic fishermen developed a similar fishery off St. Matthew Island in 1977. Landings peaked at 5,000 mt by 110 vessels from the Pribilof area in 1980-1981 and 4,300 mt by 164 vessels from the St. Matthew area in 1983. Both fisheries have been closed since 1999 owing to low abundance. Following the decline of red king crab fisheries, new fisheries developed for other species, such as the golden king crab (*Lithodes aequispinus*; see Figure 5.13c). The largest golden king crab fishery developed in the Aleutian Islands in the 1980s. Harvest from the Adak Island

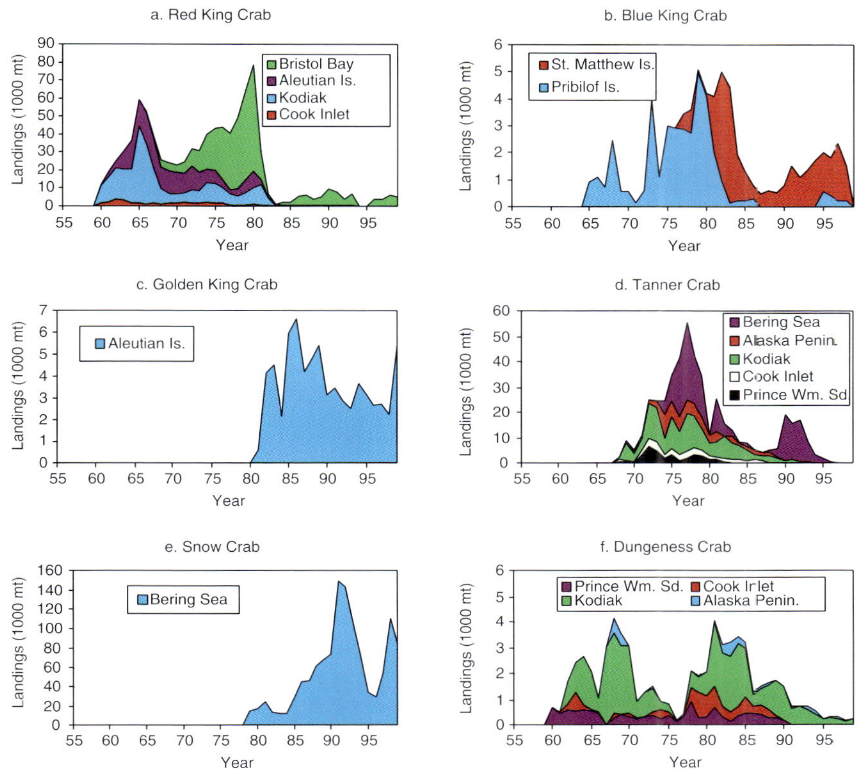


FIGURE 5.13 Historical landings from commercial fisheries for (a) red king crab, (b) blue king crab, (c) golden king crab, (d) Tanner crab, (e) snow crab, and (f) Dungeness crab from major crab fisheries conducted in the region occupied by the western stock of Steller sea lions. For Tanner crab, the Alaska Peninsula includes two fishing districts: Chignik and South Peninsula. Not shown are commercial crab fisheries in southeastern Alaska and relatively small crab fisheries for golden king crab in Prince William Sound, hair crab in the Bering Sea, and red king crab in Norton Sound.
SOURCE: Data from Shareef Siddeek, Alaska Department of Fish & Game, Juneau, personal communication (2002).

area peaked at 5,800 mt by 62 vessels in the 1981-1982 season and 900 mt by 17 vessels in 1995-1996 from the Dutch Harbor region. A small golden king crab fishery developed in Prince William Sound, peaking at just 67 mt in 1982-1983.

Fisheries developed for Tanner crab (*Chionoecetes bairdi*) in the late 1960s and early 1970s (see Figure 5.13d). The largest fishery in the Gulf of

Alaska occurred off Kodiak Island, which peaked at 13,000 to 15,000 mt in the mid- to late 1970s. The largest fishery occurred in the eastern Bering Sea, where 30,000 mt were landed in 1977 and 15,000 to 20,000 mt were caught in the early 1990s. Most Tanner crab stocks collapsed in the Gulf of Alaska in the 1980s. The eastern Bering Sea stock declined in the early 1980s but rebounded with several strong year classes that were recruited to the fishery in the late 1980s and early 1990s. In the late 1990s this stock declined again, and the fishery has been closed since 1997.

A large fishery for snow crab (*C. opilio*) developed in the late 1980s (see Figure 5.13e). This stock has gone through two large swings in abundance, with peak landings of >150,000 mt in 1992 and 1993, a sharp drop in landing to ≈30,000 mt in 1996 and 1997, and a second peak of 110,000 mt in 1999. More recently, the stock has declined sharply once again.

Fisheries developed for Dungeness crabs (*Cancer magister*) in the 1960s (see Figure 5.13f). Important fisheries occur in southeastern Alaska and Yakutat (east of 144° W). The Kodiak fishery is the largest in the central and western Gulf of Alaska. Many Dungeness crab fisheries show a cycle in landings, similar to stocks along the U.S. Pacific northwest coast. Dungeness crab stocks in Prince William Sound and Cook Inlet have not recovered from declines in the 1980s.

Small fisheries exist for other crab species as well. Korean hair crab (*Erimacrus isenbeckii*) harvests from the Bering Sea peaked at 1,100 mt in 1980. Effort ranged from 2 to 99 vessels, with less than 10 vessels in most seasons. Since the 1990s, even smaller fisheries developed for other species, including grooved Tanner crab (*Chionoecetes tanneri*), triangle Tanner crab (*C. angulatus*), and scarlet king crab (*Lithodes couesi*).

In the early years of most Alaskan crab fisheries, fishing seasons lasted much of the year. However, with increasing vessel participation and reductions in catch quotas as stocks became depleted, fishing seasons became shortened. For instance, a red king crab fishery in Bristol Bay landed 3,700 mt in just a 4-day fishing season during October 16-10, 2000. With the exception of Dungeness and golden king crab fisheries, most Alaskan crab fisheries occur in winter.

It is difficult to assess the potential role of shooting by crab fishermen on the decline of the western stock of Steller sea lions owing to a lack of data on the numbers of shot animals. Inferences must be drawn from landings data. The largest red king crab fisheries in the Gulf of Alaska and Aleutian Islands occurred in the 1960s and 1970s, before the steep decline of Steller sea lions. Large red king crab fisheries, primarily in Bristol Bay, persisted through 1982, when the sea lion decline started in the Aleutian Islands. Most red king crab fisheries have been closed since 1983. On the other hand, Tanner crab fisheries persisted into the late 1980s, and fisheries for golden king crab (mostly Aleutian Islands) and snow crab (Bering

Sea) developed in the 1980s during the sharpest period of sea lion declines. Dungeness crab fisheries (particularly in Kodiak) underwent two periods of increased landings, once in the 1960s and 1970s when sea lion abundance was high and once in the early 1980s when sea lions declined.

## Salmon Fisheries

Fisheries for salmon are considered here because salmon are among the top 10 most frequently observed prey species in the diet of Steller sea lions (Hoover, 1988; National Marine Fisheries Service, 2000) and because of historical conflicts between sea lions and salmon fisheries (see Chapter 6). The apparent importance of salmon in the diet of sea lions depends on area, season, and data collection methods. On average, from the 1950s to the 1980s salmon were found in 5.6 to 7.1% of the stomachs of the eastern stock of Steller sea lions and just 1.1 to 1.8% of those collected from the western stock (National Marine Fisheries Service, 2000, see Figure 4.5). Pitcher (1981) found salmon in an average of 3.9% of 250 sea lion stomachs sampled during 1975-1978, but their presence was virtually confined to spring and summer, with salmon in only 1% of stomachs sampled during October to March. On the other hand, salmon were present in 12% (winter) to 25% (summer) of scat samples collected in the 1990s (National Marine Fisheries Service, 2000, see Figure 4.6).

Five salmon species reside in Alaska: pink (*Oncorhynchus gorbuscha*), sockeye (*O. nerka*), chum (*O. keta*), coho (*O. kisutch*), and chinook (*O. tshawytscha*). Salmon spawn in fresh water. Embryos overwinter in bottom substrate, and fry emerge in spring. Young salmon enter saltwater after 0 to 3 years of freshwater residence, depending on species. After extensive ocean migrations for 1 to 4 years, depending on species, maturing salmon return to natal freshwater systems for spawning.

Salmon have been an important food source to Alaska natives for thousands of years. The first small salmon fishery began in Kodiak in 1785 to provide dried salmon to Russian fur traders, and the Russian American Company exported small amounts of salted salmon in the early 1800s. However, large-scale development of Alaskan salmon fisheries followed construction of the first salmon cannery in Alaska in 1878 (see Figure 5.14).

In the early 1950s, Japan developed a large high-seas salmon fishery in the North Pacific using drift gillnets. A mothership fishery peaked at 16 motherships and 506 catcher boats in 1956. Catches from this mothership fishery and a land-based fishery declined through the 1970s, and in 1988 the Japanese drift gillnet fishery ended in the U.S. EEZ and part of the Bering Sea.

Record-high commercial catches occurred in the 1980s and 1990s (see

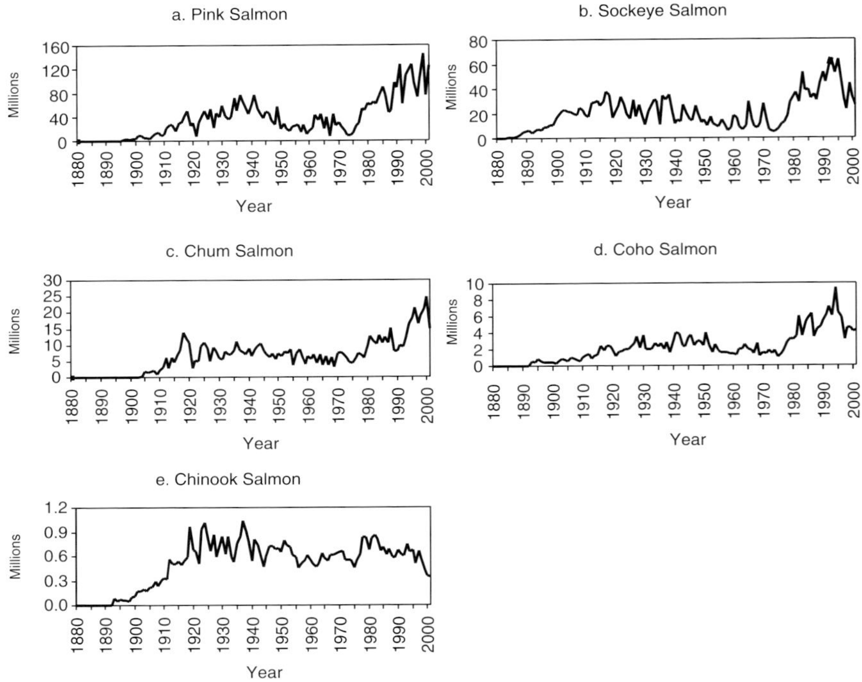


FIGURE 5.14 Historical commercial catches (millions of fish) of (a) pink, (b) sockeye, (c) chum, (d) coho, and (e) chinook salmon in Alaska. Data for 2001 are preliminary.
SOURCE: Data from Herman Savikko, Alaska Department of Fish & Game, Juneau, personal communication (2002).

Figure 5.14). However, there are exceptions, most notably chum and chinook salmon in western Alaska in recent years (Kruse, 1998). In the 1990s, on average, catches were dominated by pink salmon (60.0%), followed by sockeye (27.6%), chum (8.6%), coho (3.4%), and chinook salmon (0.4%). Most pink salmon are caught in southeastern Alaska and Prince William Sound, with significant landings off Kodiak Island. By far the largest sockeye salmon fishery occurs in Bristol Bay, but significant catches are taken from southeastern Alaska to the Alaska Peninsula. The largest fisheries for coho, chum, and chinook salmon generally occur in southeastern Alaska, but smaller fisheries exist throughout the state. Most salmon are caught by gillnet, purse seines, and troll fisheries. Fisheries are generally managed using sequences of openings and closures in attempts to allow an optimal number of adult fish to return for spawning.

As with other fisheries, it is difficult to draw definitive conclusions about potential linkages between salmon fisheries and Steller sea lion declines. Salmon catches increased in the 1980s, a period of sharpest sea lion declines. However, these increased catches were associated with large increases in salmon runs and they continued until the late 1990s, so it is not possible to associate increased catches with reduced availability of salmon to sea lions as prey. Increases in Alaska salmon production are associated with the late 1970s climate regime shift (Beamish and Bouillon, 1993; Hare and Francis, 1994). An earlier period of reduced salmon abundance corresponds to an alternate climate regime in the late 1940s to late 1970s, during which sea lion populations were much healthier. Some lethal sea lion interactions with salmon fishing gear and fishermen have been documented (see Chapter 6), and apparently some salmon fisheries continue to occur in close proximity to sea lion haulouts (Kruse et al., 2001), at least providing the opportunity for disturbance. The issue of sea lion shootings by fishermen is considered in Chapter 6.

## MANAGEMENT MEASURES TO MITIGATE POTENTIAL ADVERSE FISHING EFFECTS

A number of management measures have been implemented in attempts to mitigate potential adverse effects of fishing on Steller sea lions since they were first listed as threatened under the Endangered Species Act in 1990 (North Pacific Fishery Management Council, 1999, 2002; National Marine Fisheries Service, 2001). Additional measures were imposed after the western stock was listed as endangered in 1997. The following timeline provides a brief overview of many of these actions.

### 1990

- Shooting firearms at sea lions or within 100 yards of sea lions is prohibited. Previously, it was legal to shoot at or near sea lions, and thousands of animals are believed to have died this way.
- The number of sea lions that could be killed incidental to fishing is reduced from 1,350 to 675 animals.
- Three-mile no-entry zones are established around major rookeries west of longitude 144° W.
- The Bering Sea pollock fishery is divided into roe (or A season, January to March), and non-roe (or B season, usually September to October) fishing seasons to spread the harvest out over time.
- An ending date for the B season is set at November 1 to prevent compression of pollock fisheries and to decrease the chance of localized depletion of prey for Steller sea lions.

- Roestripping is banned statewide, partly due to concerns that discards attract sea lions to the fishing grounds.

**1991**

- Trawl closure areas are implemented to reduce disturbance of feeding sea lions within 10 nm of 27 rookeries.
- Some trawl closures are extended to 20 nm during the pollock A season.
- The domestic Bogoslof Island fishery ends due to resource conservation concerns not solely related to sea lions.

**1992**

- Trawl closure areas are implemented to reduce disturbance of feeding sea lions within 10 nm of 37 rookeries.
- Trawl closures are expanded to 20 nm around five rookeries.

**1993**

- The Atka mackerel harvest quota is apportioned among several small subareas in attempts to prevent localized depletion.

**1998**

- An improved retention and utilization program takes effect, which bans the discard of pollock and Pacific cod in all fisheries off Alaska, regardless of gear type or fishery. Discards are only permitted at times required to avoid exceeding the TAC.
- The Atka mackerel fishery is further dispersed both temporally and spatially. The TAC was equally split into two fishing seasons—January 1 to April 15 (A season) and September 1 to November 1 (B season). A 4-year schedule is implemented to reduce the percentage of harvest in sea lion critical habitat in the central and western Aleutian Islands from 80% (1999) to 40% (2002). No similar reduction plan is adopted for the eastern subarea, but 20-nm trawl exclusion zones around Seguam and Agligadak rookeries, previously in effect for pollock trawling only, are now in effect year-round for the Atka mackerel fishery.
- In the Gulf of Alaska, full retention is required for Pacific cod.
- Fishing is banned for forage fishes, such as capelin, sand lance, and others in efforts to prevent the development of new fisheries that

could compete with higher trophic levels, such as groundfish, seabirds, and marine mammals.

### 1999

- Trip limits are imposed for the pollock fishery in the Gulf of Alaska.
- Pollock fishing in the Aleutian Islands is prohibited.
- Pollock catches in critical habitat areas are restricted.
- Four fishing seasons are created for pollock in attempts to further spread the harvest out over time.

### 2000

- More rookeries are added to the trawl closures.
- Several new measures are instituted to spread out the catch of Pacific cod throughout the year.

Obviously, the potential need for additional fishery restrictions depends on the chances that fishing adversely affects Steller sea lions and the degree to which these effects may have been mitigated by measures already implemented. Because there have been frequent changes in sea lion protective measures since 1990, it is very difficult to evaluate the impacts of individual management measures on fish distribution, Steller sea lions, and the fishery.

# 6

# Steller Sea Lion Decline: Environmental Context and Compendium of Evidence

Evaluation of the main hypotheses proposed for the causes of decline and failure to recover of the western Steller sea lion population depends on understanding how food web linkages affect sea lions. Every species in an ecosystem, including humans, contributes to and is affected by the linkages represented in a food web. Food webs are described by identifying who eats whom, based on direct observation, stomach and scat analyses, or prey item chemical signatures based on stable isotope and fatty acid analyses. Understanding how linkages influence population and ecosystem dynamics is a far greater challenge because the complexity of interactions precludes analysis through static observation. Only through perturbation of one or more populations is it possible to evaluate the dynamic nature of food webs. Though much is known about the descriptive structure of food webs, the dynamic properties are less well understood. Humans are part of the food web; in the current case, they may change food web dynamics through direct takes of sea lions, removal of sea lions' preferred prey, removal of alternate prey items of sea lion predators, or some combination of the above. For example, humans depleted populations of several whale species, possibly inducing killer whales to increase their predation on sea lions, seals, and otters.

This chapter describes four concepts that provide a context for analyzing the role of food web interactions in Steller sea lion population dynamics and then applies these concepts to evaluate the many hypotheses proposed to explain the Steller sea lion population decline.

## FOOD WEB CONCEPTS

### Bottom-Up and Top-Down Control

Food web linkages connect species across different trophic levels (Paine, 1980). The functional significance of any linkage can be viewed from the perspective of either the consumer or the prey. When a population's size is limited by the availability of prey, it is described as *bottom-up control*; when the size of a population is determined by predation, it is described as under *top-down control*. Bottom-up control characterizes populations that decline or fail to expand because there is insufficient food for growth, reproduction, or survival. Top-down control characterizes populations whose size is regulated by the abundance and feeding habitats of the species that prey on them.

### Direct and Indirect Food Web Linkages

Food web linkages can connect two species directly or indirectly if there are one or more intermediate species. Although hypotheses based on direct effects could be sufficient to explain the decline of Steller sea lion populations, indirect effects may influence the pattern of decline and thereby complicate analysis of direct impacts (see Box 6.1). How the number of intermediate species affects the strength of the interaction is generally unknown, but the number of potential indirect linkages is far greater than the number of direct linkages. Therefore, it is insufficient to consider only the availability of specific prey items or the foraging patterns of generalist predators because changes among other members of the food web may indirectly affect sea lion survival.

Humans have exploited many large and small predators in marine ecosystems, acting as an agent for *top-down control* of marine populations. If top-down forcing is important, the depletion of apex predators should have strong effects on food webs through increased abundance of species at lower trophic levels; removal of predators disrupts the trophic cascade (Paine, 1980; Carpenter and Kitchell, 1993). The consequences of these shifts in marine food webs are complex, difficult to predict, and often unrecognized.

### Scale and Connectivity

Most analyses of food web dynamics have focused on the linkages among species in a common ecosystem. There is growing evidence for the importance of linkages across ecosystems. Linkages of this sort occur in a variety of forms and connect otherwise functionally distinct ecosystems

## BOX 6.1
## Some Plausible Examples of Indirect Effects

The Bowen et al. (2001) report presented a number of hypotheses concerning the history of Steller sea lion population trends since 1960 to identify the most informative parameters for guiding future research programs. These hypotheses address direct effects that may explain the decline in Steller sea lions. For example, the discussion of trophic consequences of fishing has focused on reduced food availability. A broader classification of "fishing effects" would also include indirect influences of diverse trophic interactions (Wootton, 1994), estimated to be approximately 50% of all ecological interactions (Schoener, 1993; Menge, 1995).

One example, captured in the following quotation from Loughlin and York (2000), is the implication that factory trawlers, simply by their presence and activity, can aggregate sea lions and killer whales. "Predation [by killer whales] is often focused in small areas, i.e., where sea lions are localized near large fish processing vessels, resulting in exacerbation of local declines" (p.43). As a second plausible example, bycatch (nontarget animals caught, killed, or injured during fishing operations) could provide a new source of food for bottom-dwelling organisms, including various flat fish. Arrowtooth flounder had several strong recruitment years in the 1980s; biomass peaked in 1995 (five-fold increase over 1980 biomass) and by the 2001 estimate had declined about 20% due to lower recruitment levels in the 1990s (Wilderbuer and Sample, 2001). The years of high recruitment occurred when the pollock fishery used bottom trawling gear that causes incidental mortality of benthic organisms (National Research Council, 2002). The high bycatch of that fishery may have contributed to the resurgence of arrowtooth flounder. These fish may pose a competitive threat to sea lions because they also prey on young pollock, other demersal fish, and invertebrates.

In a third indirect effect scenario, the 10-fold increase in jellyfish over the past decade could also deplete the sea lions preferred prey (Brodeur et al., 1999). Plausible interaction pathways could include direct competition with age 0 pollock for zooplankton or predation on the smaller pollock. This massive increase in jellyfish might constrain the recruitment of many commercially and ecologically significant fish stocks. In the Black Sea, studies have documented the top-down control of commercially valuable fish stocks by a comb jelly (Shiganova, 1998). Further sampling and analysis of these food web linkages in the Bering Sea would be needed to fully evaluate the impacts of this increase in jellyfish on recruitment of important sea lion prey species.

Indirect effects of fishing activity—especially if concentrated seasonally and spatially—could account for some of the unexplained mortality in the western population of Steller sea lions. In any event, despite the difficult challenge of unraveling their impacts, indirect effects should not be dismissed as either biologically unusual or dynamically trivial.

over a wide range of spatial and temporal scales. Many marine species have dispersive life stages that can be carried great distances by ocean currents. Also, large animal movements can link disparate ecosystems in important ways. The altered foraging behavior of killer whales (to include sea otters in their diet) provides a link between the kelp forest ecosystems of the Aleutian archipelago and the food web of the open ocean (Estes et al., 1998).

Previous shifts in the abundance of key species continue to affect the dynamics of present-day food webs. For instance, the progressive removal of herbivorous fishes and invertebrates by historical fisheries from Caribbean coral reefs in conjunction with a mass die-off of sea urchins was likely responsible for declines in coral abundance due to overgrowth by algae (Jackson et al., 2001). Thus, it is possible that the Steller sea lion population decline and failure to recover is in part influenced by events distant in either time or space.

### Alternative Stable States

Because food webs have complex multiple linkages, the response of these systems to disturbance is often nonlinear (Ruesink, 1998). For instance, if a major predator is removed, the increased availability of prey resources may allow expansion of other predatory species to a new, relatively stable equilibrium. This alternative stable state, dominated by a different assemblage of species, may inhibit the return of the food web to its previous status (Lewontin, 1969; Holling, 1973; May, 1977; Sheffer et al., 2001). This concept has important implications for management because of the possibility that disturbed ecosystems may not return to their previous state of equilibrium. Hence, even if the causes of the Steller sea lion decline are identified and addressed, the western sea lion population still may fail to reach its former abundance.

## MULTIPLE WORKING HYPOTHESES

At least eight hypotheses have been proposed to explain the rapid decline of the western stock of Steller sea lions. As pointed out in Box 1.1 (Chapter 1), these various hypotheses cannot be accepted or rejected through the method of strong inference. The data necessary to conduct determinative analyses were simply not collected during the years of the rapid decline. However, this is not to say that relevant data are entirely lacking, particularly with regard to current trends in the population. Numerous types of information on Steller sea lions and their environment have been obtained over the years, some fortuitously and some for the specific purpose of trying to better understand sea lion ecology and population biology. Although none of this information is sufficient to prove or eliminate hypotheses, much of it can be rated according to its consistency with any given hypothesis. When all of the information is assessed in aggregate, a weight of evidence argument emerges that allows ranking of the hypotheses according to conformity with available information.

The main hypotheses that have been proposed to explain the Steller sea lion decline are described in Table 6.1. Each hypothesis is presented separately for the sake of clarity, but this should not be taken to imply that the hypotheses necessarily act independently of each other nor does it preclude the possibility that the recent decline results from a combination of the hypothesized causes.

## FOOD LIMITATION—BOTTOM-UP HYPOTHESES

Under the bottom-up scenario, the Steller sea lion decline is attributed to a deficiency in food resources. This deficiency could be manifested as depletion of prey, reduced abundance of preferred prey species, or reduced accessibility to prey due to local depletion or disturbance of fish stocks. Nutritional limitation caused by either a climate regime shift and/or a fisheries effect requires that either the quantity or quality of food is insufficient for the recovery or maintenance of the Steller sea lion population. This could come about from starvation conditions, nutritional impacts on reproductive success, or increasing susceptibility of animals to disease.

During the period of rapid decline in the 1980s, the demographics of the western stock gave some indications that Steller sea lions were nutritionally stressed. In 1985, sea lions were on average smaller, were slower to reach reproductive maturity, and had a lower birth rate than in the 1970s (Calkins and Goodwin, 1988; York, 1994). There was also evidence of higher rates of abortion and lower juvenile survival (Pitcher et al., 1998). Nutritional stress may have been a contributing factor in causing the rapid decline of the western population of sea lions, but models indicate that reduced prey availability *alone* is unlikely to account for the dramatic decline in the size of the population (see Chapter 3).

In 1991 the Alaska Sea Grant College Program (1993) sponsored an international conference entitled "Is It Food?" to ascertain what kind of physiological or biochemical changes would be expected in a chronically or acutely food-stressed pinniped. Because of the difficulty in handling Steller sea lions, the first field studies to address these questions were conducted during the summer on newborn pups and adult females on rookeries. They utilized an east versus west comparative approach with the hypothesis that the declining western population would be stressed relative to the stable eastern population. The studies looked at pup growth

TABLE 6.1 Eight Major Hypotheses Proposed to Explain the Steller Sea Lion Population Decline. Each hypothesis is characterized by purported demographic mechanism(s) of population change, food web forcing directions, and the acronyms used later in Table 6.2. Although the cause of the sea lion decline likely falls within this breadth of hypotheses, more than one of the listed hypotheses may have contributed to the decline, additively, interactively, or in various degrees of relative importance in different places or at different times.

| Hypothesis | Mechanism of Population Limitation | Forcing Direction | Acronym |
|---|---|---|---|
| 1. Fisheries removal | Starvation and/or reproductive failure because of nutritional limitation | Bottom-up | FR |
| 2. Climate change/ regime shift | Starvation and/or reproductive failure because of nutritional limitation | Bottom-up | CE |
| 3. Predation | Elevated mortality from attack by predators | Top-down | PRED |
| 4. Direct take | Elevated mortality from shooting or other purposeful killing | Top-down | DT |
| 5. Subsistence harvest | Elevated mortality from shooting for food or other subsistence uses of sea lions | Top-down | SH |
| 6. Incidental take/ entanglement | Elevated mortality from entanglement in fishing gear due to injury or drowning | Top-down | IT/ENT |
| 7. Disease | Elevated mortality or reproductive failure caused by parasites, viruses, or bacteria | Top-down | D |
| 8. Pollution/ biotoxins | Elevated mortality or reproductive failure from poisonous or toxic substances, either natural or human produced | Top-down or Bottom-up | PO |

rates, maternal attendance patterns, blood chemistry profiles, milk quality, at-sea metabolic rate estimates, thermoregulatory measurements, and several other variables that had been outlined in the "Is It Food?" conference. These studies in the mid-1990s found that animals in the western population were at least as healthy as in the southeastern Alaskan populations based on several measurements of body condition such as birth size, pup growth, and adult size. This research was summarized in the "Is It Food? II" conference in 2001 (see DeMaster and Atkinson, 2002).

Conclusions based on these results are limited because of sample size (less than 20 adult females and less than 100 pups), seasonality (they were only conducted in the summer on rookeries), and insensitivity to subtle differences between populations. Despite these limitations, the studies suggest that it is unlikely that newborn pup survival has been compromised by acute or chronic malnutrition over the past decade. These studies have now been expanded to juveniles on a year-round basis because new capture methods allow large numbers of juveniles to be handled. All preliminary evidence shows similar results: sea lions in the western population show no indication of being nutritionally stressed relative to sea lions in the eastern population (Richmond and Rea, 2001). The consensus statement drafted from the "Is It Food? II" conference (Alaska Sea Grant College Program, 1993) states that nutritional limitation is probably not a major contributor to the population decline over the past 10 years. Additional studies on animals from a variety of locations would be necessary to establish whether these results apply generally to Steller sea lions throughout the western range.

The following sections describe the two mechanisms proposed to cause a decrease in the availability or quality of the food supply for Steller sea lions throughout the history of the decline. These two mechanisms, climate regime shifts and fishery removals, may have had a combined effect that limited the availability of common Steller sea lion prey items during the earlier phases of the decline.

### Climate Regime Shift

The regime shift hypothesis links climate-forced environmental changes to changes in the welfare of Steller sea lions through indirect trophic interactions in the marine food web. Several different mechanisms have been proposed that link climatic regime shifts to declines in Steller sea lions in the 1970s and 1980s. A reduction in the abundance of herring, capelin, and sand lance and a concomitant increase in large piscivorous fish associated with the 1977 climatic regime shift (see Chapter 2) may have adversely affected Steller sea lions by reducing the proportion of high-calorie fish in their diet (see discussion of the junk food hypothesis below). Merrick et al. (1997) showed that declines in sea lion populations correlate with a decrease in sea lion dietary diversity, which may be indicative of a change in the availability of prey species (Anderson and Piatt, 1999).

The regime shift hypothesis largely rests on statistical inference, wherein correlation analyses identify statistically significant associations between many 20th-century climate, fishery, and ecosystem survey records across the broad geography of the North Pacific and Bering Sea (e.g., see Anderson and Piatt, 1999; Hare and Mantua, 2000). Potential mechanisms linking climate changes to ecosystem changes in the North Pacific and Bering Sea are reviewed in Chapter 2. Testing the regime shift hypothesis is essentially limited to a "wait and see" approach that cannot distinguish between the impacts of natural environmental changes and other perturbations. Based on a recent shift to cooler upper-ocean temperatures and a weakening of the wintertime Aleutian Low beginning in 1998, the climate may have shifted to a "cool phase" Pacific Decadal Oscillation (PDO) state similar to what existed before the steep decline in Steller sea lion populations (Hare and Mantua, 2000; Schwing and Moore, 2000; Peterson and Mackas, 2001). If PDO regime shifts exert a significant and reversible forcing on sea lion abundance, the western population should begin to recover in response to the 1998-2002 climate trends.

## Fishery Removals

The spatial and temporal scales of commercial fisheries provide important insights for evaluating the possible effects of fishery removals on the nutritional status of Steller sea lions. Fisheries and their potential interactions with Steller sea lions were discussed in detail in Chapter 5. This section first evaluates whether fisheries have depleted prey resources at a regional scale on an interannual basis to the extent that there is insufficient fish biomass to sustain the extant number of sea lions. The second part of this section considers the potential for fisheries to deplete sea lion prey at a local scale.

### Evidence for Broad-Scale Depletion

Fishery and stock assessments indicate that walleye pollock (except the Donut Hole stock), Pacific cod, and Atka mackerel stocks are not overfished (North Pacific Fishery Management Council, 2001a, 2001b). Periodic strong year classes drive much of the change in fish stock abundance. Although parental abundance affects recruitment at low stock levels, some if not most of the recruitment variability in groundfish stocks in the North Pacific Ocean appears to be associated with variability in environmental conditions. Regime shifts toward winters with a deepened Aleutian Low Pressure System tend to be associated with higher frequencies of strong year classes among groundfish stocks (Hollowed and Wooster, 1992, 1995).

Are there enough fish to support a healthy population of Steller sea lions? This is a difficult question to answer without making many assumptions, but it addresses only the simplest consequence of the fishery. The more comprehensive question is whether the appropriate species, sizes, and densities of prey are available at spatial and temporal scales necessary for foraging sea lions. As will be shown, this is also a difficult question to answer given the poor state of knowledge about sea lion foraging ecology, fine-scale distributions of fishes, and effects of fishing on fish school dynamics.

In the 2000 Biological Opinion (BiOp #3), the National Marine Fisheries Service attempted to determine whether there was sufficient groundfish prey for Steller sea lions by calculating the amount of food consumed by sea lions relative to the biomass of the groundfish in the Gulf of Alaska, Aleutian Islands, and Bering Sea in 1999. This yielded a ratio of biomass consumption to availability of 1:54. Similarly, the agency deduced that a historical high number of 184,000 Steller sea lions would consume about 1.7 million metric tons (mt) annually, for a ratio of consumption to availability of 1:21. This comparison indicated that the current availability of fish biomass is higher for the 1999 population than for the prefishery population of sea lions.

Another approach for examining simple food availability is to compare trends in pollock, Pacific cod, and Atka mackerel biomass with Steller sea lion counts. Figure 6.1 shows estimates of exploitable fish biomass. Steller sea lion counts correspond to index sites in the Gulf of Alaska and Aleutian Islands, as there are no index rookeries in the Bering Sea. Also, analyses of sea lion counts were restricted to years in which at least 24 or 35 index rookeries were observed in the Gulf of Alaska and Aleutian Islands, respectively; counts from years with fewer sites tended to be inconsistent with counts from adjacent years.

In the Gulf of Alaska, sea lion numbers declined from the 1950s through the 1970s, a period during which pollock abundance was increasing (Figure 6.1a). The most rapid decline of sea lions occurred from 1977 to 1985, when pollock landings peaked in the Gulf of Alaska (Figure 5.2) and there was a large increase in the percentage of groundfish taken in Steller sea lion critical habitats (Figures 5.10 and 5.11). Trawl surveys in 1984-1996 did not show any decrease in high-density pollock abundance. However, many sea lions were taken as bycatch in the fishery. During this same time period in the 1980s, Pacific cod abundance increased (Figure 6.1c). Abundance of all three species declined from the late 1980s to the present. However, regardless of whether one considers pollock biomass alone or combined pollock and cod biomass, there have been more of these fish species available per Steller sea lion since the mid-1980s than prior to 1980 (see Figure 6.2a).

In the Bering Sea and Aleutian Islands, Steller sea lion counts were high from the mid-1960s to the late 1970s, at a time when eastern Bering Sea pollock abundance appears to have been low to moderate (Figure 6.1b). Steller sea lions declined sharply in the 1980s at a time when pollock, cod, and Atka mackerel abundances were increasing (Figure 6.1b, d, e). Interestingly, pollock biomass apparently declined during 1989-1992, whereas the Steller sea lion decline abated during 1990-1992 in the Aleutian Islands. As with the Gulf of Alaska, the biomass of pollock and the combined biomass of pollock, cod, and Atka mackerel per sea lion were higher after 1985 than prior to 1980 (Figure 6.2b).

As mentioned earlier in Chapter 5, these comparisons are fraught with assumptions and therefore should be interpreted with caution. Several factors tend to lead to overestimation of the fish biomass available to Steller sea lions. For example, in this analysis the committee chose index sites because this subset of rookeries provides the best measure of trends over time. However, by definition, these indices underestimate the total abundance of Steller sea lions because they do not include counted animals on nonindex sites and uncounted animals at sea during the surveys. Additionally, only index sites in the Gulf of Alaska within the range of the western stock of Steller sea lions were considered. Increases in the abundance of Steller sea lions in southeastern Alaska (eastern stock) were not included, although fish biomass estimates are gulf-wide values. Moreover, pollock in the northwestern portion of the eastern Bering Sea may not be available to sea lions, but the values represent total exploitable pollock biomass over the entire continental shelf.

Other assumptions of this analysis contribute to underestimation of prey biomass available to Steller sea lions. For instance, biomass estimates of Aleutian Islands pollock were not included because of questions about the discreteness of this stock. In 2000 a bottom-trawl survey estimated 105,500 mt of pollock in the Aleutian Islands (Ianelli et al., 2001). Survey trends indicate that pollock abundance in this area peaked in 1983, declined until 1994, and increased since then. Likewise, pollock biomass in the Aleutian Basin and Bogoslof Island areas were not considered in our analysis. It also should be noted that Steller sea lions target juvenile pollock (Merrick and Calkins, 1996), yet young pollock are not fully sampled by the survey gear and tend to be underestimated in stock assessments. Finally, our estimates do not include other components of the sea lion diet, including fish (e.g., salmon, herring, flatfishes, rockfishes, sand lance, capelin) and invertebrates (e.g., octopus, squid). Given these caveats, this analysis does not provide support for the hypothesis that the recent decline in the Steller sea lion population is due to depletion of sea lion prey by the groundfish fisheries.

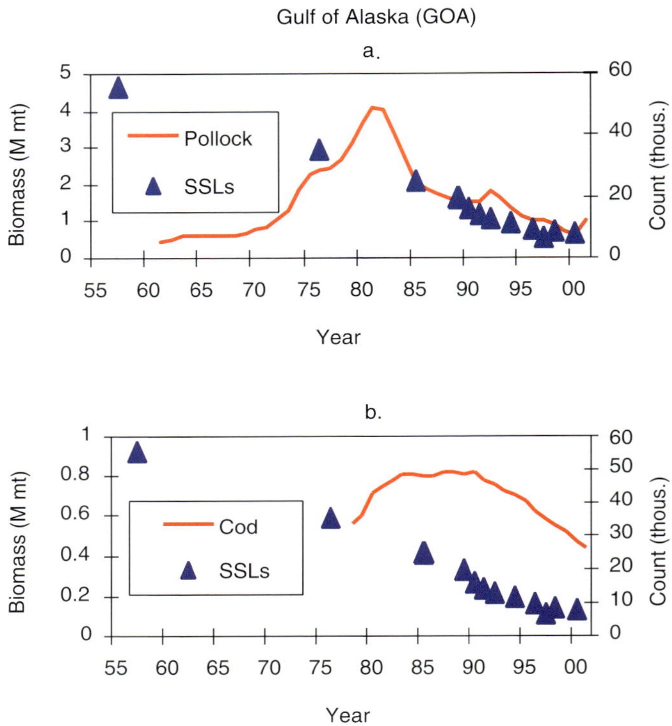


FIGURE 6.1 Trends of Steller sea lion (SSL) index counts in the Gulf of Alaska (GOA) versus exploitable biomass of (a) GOA pollock and (b) GOA Pacific cod. Trends of SSL index counts in the Aleutian Islands versus exploitable biomass of (c) Bering Sea/Aleutian Islands (BS/AI) pollock, (d) BS/AI Pacific cod, and (e) BS/AI Atka mackerel. Index counts are for those years in which a minimum of 24 rookeries in the Gulf of Alaska or 35 rookeries in the Aleutian Islands were observed.
SOURCE: Data from National Marine Fisheries Service, National Marine Mammal Laboratory, Seattle, available at *www.afsc.noaa.gov/*.

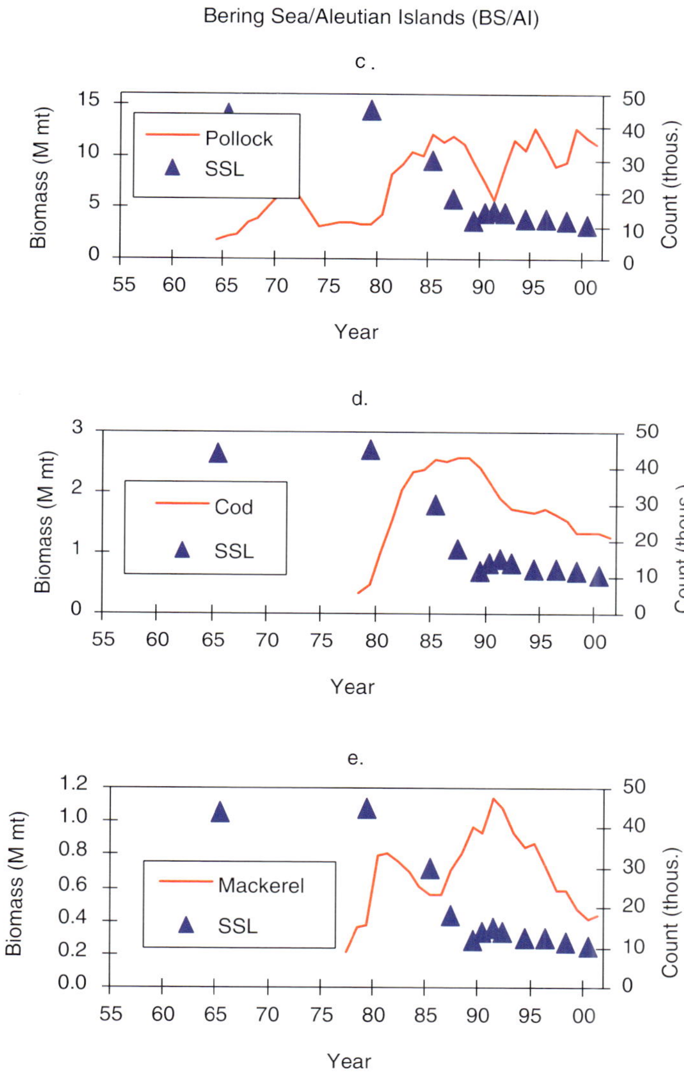


FIGURE 6.1 Continued from previous page.

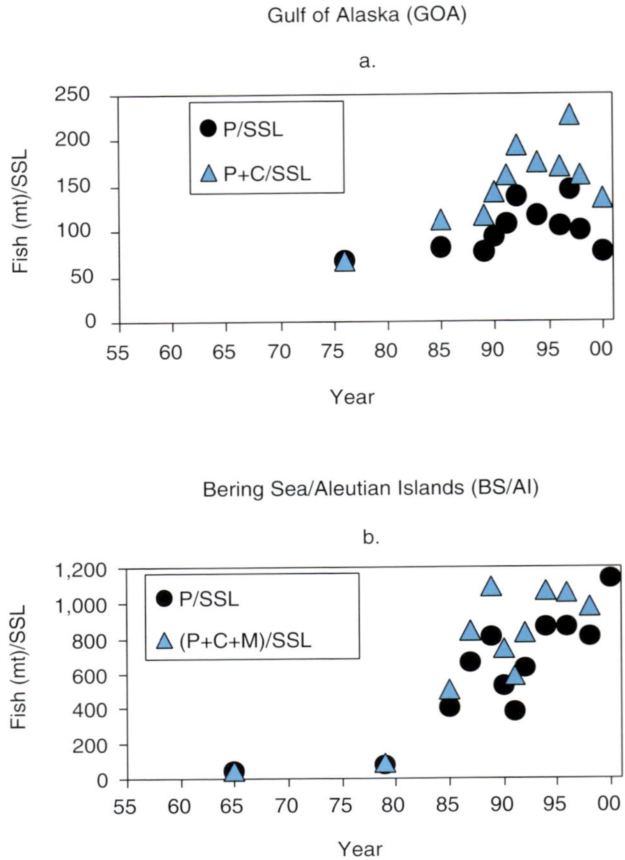


FIGURE 6.2 Relative availability of fish biomass of pollock (P), pollock and cod (P+C), or pollock, cod, and mackerel (P+C+M) per Steller sea lion (SSL) index count for (a) Gulf of Alaska and (b) Bering Sea/Aleutian Islands. Fish biomass estimates are based on stock assessments conducted for 2001.
SOURCES: Data from Dorn et al. (2001); Ianelli et al. (2001); Lowe et al. (2001); Thompson and Dorn (2001); Thompson et al. (2001).

## Evidence for Localized Depletion

Although the North Pacific groundfish are managed conservatively based on single-species harvest strategies, concerns remain over other ecological effects, such as localized depletion in the vicinity of sea lion rookeries, have remained (National Marine Fisheries Service, 2000). As described in Chapter 5, few studies have been conducted on localized depletion of sea lion prey by commercial fisheries. There is evidence of localized depletion of Atka mackerel in some areas in the early to mid-1990s (Fritz, 1999). Atka mackerel tend to stay in a given area, rendering this species particularly susceptible to local depletion. The recovery of biomass to prefishery levels by the start of the subsequent fishing season indicates that depletion is temporary, but effects of this seasonal depletion on foraging Steller sea lions are unknown.

Preliminary results from an ongoing study of pollock suggest that trawl fisheries in the Barnabas trough off Kodiak Island do not cause localized depletion (Wilson et al., 2002). The biomass and distribution of pollock were relatively stable over periods of days to weeks. Also, there was no discernable effect on the vertical distribution or school fractal dimensions. These preliminary results, and reports of high-density pollock areas in the Gulf of Alaska during 1984-1996 (Shima et al., 2002), indicate that localized depletion of pollock is unlikely. Additional research is needed on other species and in other areas before concerns can be fully resolved about localized depletion by fishery removals.

### Summary of Evidence for Fishery Removal Hypothesis

Under the fishery removal hypothesis, fisheries may have sufficiently depleted the availability of prey for Steller sea lions, causing nutritional stress with adverse ramifications on reproduction and survival. Owing to the lack of studies specifically designed to test this hypothesis, it cannot be accepted or rejected. Data derived from fish stock assessments and commercial fisheries indicate that the overall levels of fish biomass in the federal fishery management areas are large relative to the current population of Steller sea lions. However, additional research on seasonal fish movements and migrations, effects of fishing on fish school size and distribution, and sea lion foraging ecology are needed to more fully resolve the issue of localized depletion.

## FOOD QUALITY

The junk food hypothesis posits that Steller sea lions are consuming a lower-quality diet (i.e., eating fewer fatty fish such as capelin and herring and more lean fish such as pollock) because of an ecosystem-level change in their environment. The most recent publication on this matter (Thomas and Thorne, 2001) suggested that sea lions cannot find enough herring in their home range and therefore eat pollock instead. The diet diversity hypothesis (Merrick et al., 1997) is a corollary of this idea; it postulates that sea lions require a more diverse diet to thrive and correlates low

**BOX 6.2**
**Diet Diversity**

Analyses of Steller sea lion scats revealed a strong inverse relationship between dietary diversity and the estimated depth of population decline (Merrick et al., 1997). Those populations that had declined the most also had the least diverse diets, and in keeping with this pattern, dietary diversity was greatest for the increasing sea lion population in southeastern Alaska. In some areas the less diverse diet consisted mostly of pollock, apparently supporting the junk food hypothesis that pollock are of insufficient nutritional quality to maintain a healthy sea lion. Similar patterns were observed in the central and western Aleutians, although there the dominant prey type was Atka mackerel. However, the inverse relationship between dietary diversity and the degree to which food resources are limiting to populations of consumers could indicate that dietary diversity in Steller sea lions decreases when prey resources are not limiting. A positive relationship between dietary diversity and the degree of food resource limitation is more consistent with foraging theory (Emlen and Emlen, 1975; Krebs, 1978). Many foraging models predict that when prey resources are abundant, consumers will specialize on the most beneficial of these. As consumer populations increase and the most beneficial or valuable prey are depleted, increasingly less valuable prey are added to the diet, thus increasing dietary diversity.

dietary diversity with declining populations (see Box 6.2). On a less optimal diet, it is assumed that sea lions would be unable to eat a sufficient amount of food to meet their metabolic demands.

Development of a metabolic cost analysis for Steller sea lions requires knowing a suite of factors that can generally be divided into the costs of thermoregulation, swimming and diving (i.e., hydrodynamics), basal metabolic demands (i.e., digestion, growth), and reproduction (pregnancy and lactation for females, harem maintenance for males). Although all of these values can be modeled or estimated based on body size or other mammal species, unless they are actually measured on Steller sea lions, it will be not be possible to know if Steller sea lions balance the overall equation of demand and supply similarly to other species. Given the potentially subtle or chronic nature of the nutritional stress, modeling efforts may not be able to provide the refinement of data necessary to determine if sea lions are food limited.

Because the diets of animals cannot be controlled in the wild, there is no direct way to test the junk food hypothesis in the field. Therefore, several laboratory-based studies have examined how Steller sea lions digest and process a variety of different fishes. Studies of digestive efficiency, gut passage rates, otolith recovery conditions, metabolic rate control, and swimming energetics are being conducted on captive animals. The first study, based on short-term feeding trials, showed that juvenile Steller sea lions could not maintain their body mass on a diet of pollock relative to herring (Rosen and Trites, 2000). More recent analysis of these data suggests that seasonal effects may have influenced the results and may only apply to juveniles, not adults (Trites, 2001).

Ongoing research programs at the Alaska SeaLife Center (ASLC) in Seward have been conducting long-term feeding trials (3 years) with young adult Steller sea lions. Preliminary results indicate that sea lions show seasonal and gender-specific responses to dietary change and maintain body condition under different dietary regimes (Castellini et al., 2001). In similar long-term studies with harbor seals in Alaska, body condition could not be predicted based on a diet of herring or pollock (Castellini et al., 2000). Current studies examine prey items in order to measure and assess their nutritional quality (calories, vitamins, lipid levels, protein levels, iron content, etc.) by both region and season. For example, sea lions in the feeding trials at the ASLC are being fed diets representative of predecline, postdecline, and southeastern Alaska conditions. Ongoing studies will provide a nutritional assessment of these diets. As of 2002, ongoing scat studies suggest regional and seasonal differences in the prey items found in sea lion scats (Sinclair and Zeppelin, 2002) and shifts in dietary diversity correlate with Steller sea lion population trends. However, this correlation may be coincidental rather than an indication of a causal link between diet and population decline.

Bowen et al. (2001) challenge the diet diversity theory because a link between diversity of the Steller sea lion diet and their lack of recovery has not been established. The Alaska Steller Sea Lion Recovery Team (Kruse et al., 2001) looked at the same evidence from Merrick as reported in BiOp #3 and recommended further consideration of the potential effect of diet diversity on Steller sea lion population trends.

In summary, the field-based physiology and feeding ecology studies have focused on defining how animals in different regions or at different ages compare with predicted responses of a nutritionally stressed population, while the laboratory-based programs have tried to quantify the metabolic pathways and control between prey quality and Steller sea lion condition. To date, evidence from these studies does not support the hypothesis that food limitation is currently reducing the condition of adult females, or impairing their ability to raise a pup, or compromising the health of animals up to 2 years of age.

## PREDATION AND TAKES BY HUMANS— TOP-DOWN HYPOTHESES

Population growth rates decline fastest when adult survival is decreased (see Chapter 3). Therefore, factors that increase adult mortality are expected to have the greatest proportional impact on the population. Both natural (predation by killer whales or sharks, infectious disease, or toxins) and anthropogenic (deliberate kills and incidental takes) sources of adult mortality could have dramatic impacts on the demographics of the Steller sea lion population.

Previous estimates of these various sources of adult mortality were too low to explain the rapid, range-wide decline in the western population (Chapter 3). However, predation by killer whales may have been higher than previously assumed, and there is substantial uncertainty in the estimates of deliberate and incidental takes by humans. Relatively small increases in either the natural or anthropogenic mortality parameters in the Ecosim model described in Chapter 3 could improve the fit of the model to the observed pattern of sea lion decline. Additionally, these sources of mortality could readily explain the continuing, albeit slower, decline of the sea lion population. These top-down mechanisms are also consistent with recent studies that have found no evidence of reduced health or fitness in comparisons of declining and increasing populations of sea lions.

### Predation on Steller Sea Lions (Killer Whales and Sharks)

Ecology has been slow to consider the role of predation in population regulation and food web dynamics, despite early examples (Elton, 1927; Brooks and Dodson, 1965; Paine, 1966) and a well-known conceptual framework (Hairston et al., 1960). But there is growing evidence from diverse ecosystems for the importance of predation and top-down forcing (Estes et al., 2001). The earliest and most compelling evidence comes from aquatic systems: lakes (Carpenter and Kitchell, 1993), streams (Power, 1990), and coastal environs (Paine, 1966; Estes and Palmisano, 1974). However, there are few known examples from oceanic ecosystems, despite the remarkable diversity and abundance of large predators in the sea. Studies of pattern and process in the ocean have been dominated by a nearly exclusive focus on bottom-up forcing mechanisms.

Evidence for predation can be evaluated based on the following three questions: Who are the potential predators? Are they sufficiently abundant and wide ranging to explain the decline based on what is known of their diets and nutritional requirements? Do current or historical records provide evidence for predator-based shifts in the sea lion population?

There are currently three contenders for the predators: salmon sharks (*Lamna ditropis*), Pacific sleeper sharks (*Somniosus pacificus*), and killer whales (*Orcinus orca*). Both shark species are abundant in the western Gulf of Alaska and their population numbers appear to have increased considerably in the 1990s (Lee Hulbert, National Marine Fisheries Service, Auke Bay Laboratory, personal communication, 2002, available at *http://fakr.noaa.gov/oil/sharks.htm*). Salmon sharks are large, fast-swimming, aggressive predators. Although related shark species attack and consume pinnipeds, there are no such records for salmon sharks, which appear to be largely or exclusively piscivorous. Sleeper shark stomachs sometimes contain remains of harbor seals and small cetaceans (Bigelow and Schroeder, 1948; Bright, 1959; Lee Hulbert, National Marine Fisheries Service, Auke Bay Laboratory, personal communication, 2002), but as yet there is no evidence they attack Steller sea lions. Also, there are no reports of shark attacks or shark wounds on sea lions. Finally, because large sharks are rarely if ever observed, it is doubtful that they could occur in sufficient abundance to account for declining Steller sea lion populations in the Aleutian Islands and southeastern Bering Sea.

The third candidate predator, the killer whale, attacks and consumes a diverse array of marine mammals and fish, including many species of large and small cetaceans, pinnipeds, and sea otters (Matkin et al., 1999). Considerable dietary variation exists among groups or individual killer whales (Ford et al., 1998). Some killer whales appear to feed largely or exclusively on marine mammals, described by some researchers as transients (Ford and Ellis, 1999). Killer whale attacks on Steller sea lions have been reported, and sea lion remains were found in the stomach contents of two of eight killer whale carcasses recovered in Alaska (Barret-Lennard et al., 1995). One of these two killer whales, a beach-cast female recovered in Prince William Sound in 1992, contained flipper tags from 14 Steller sea lions tagged as pups on Marmot Island in 1987-1988 (Saulitis et al., 2000). Killer whales have been implicated in the collapse of sea otter populations in coastal waters across the Aleutian archipelago and northern Gulf of Alaska (Estes et al., 1998). Theoretically, a switch of fewer than four killer whales to feeding exclusively on sea otters can account for the additional annual mortality in the central Aleutian Islands during the rapid decline of the sea otter population. Like the sea otter, sea lion declines could be explained by remarkably small changes in killer whale foraging behavior based on the energetic requirements of killer whales. Scientific surveys of killer whale abundance have only recently been conducted, so the historical size of the killer whale population in Alaska is unknown. However, the killer whale population in the southeastern Bering Sea has recently been estimated at 391 (95% confidence interval = 171-894), and this area represents only a fraction of the range of the western population of Steller sea lions (Waite et al., 2002).

If killer whale predation on Steller sea lions increased, what might have triggered this change? Possible explanations include an increased abundance of predators, a change in feeding behavior, or a decrease in the availability of other marine mammal prey species. Existing information is insufficient to answer this fundamental question. The North Pacific and Bering Sea ecosystems have experienced a number of significant perturbations, including the depletion of great whales between the late 1940s and early 1970s, the establishment of large commercial fisheries, and changes in the oceanic food web caused by a climate regime shift in the late 1970s. These or possibly other factors may have influenced the population dynamics and foraging behavior of the ecosystem's apex predators.

Because killer whales must consume considerable amounts of prey (in terms of weight) to meet their metabolic requirements, a relatively small change in the rate of killer whale predation on Steller sea lions could account for the historical decline. Although Branson (1971) reported that attacks by killer whales on pinnipeds were unknown prior to about the 1970s, there is no direct evidence that increased predation since then was the primary cause of Steller sea lion decline in the 1980s. Hence, the role of killer whale predation in the historical decline is indeterminate, but further research on killer whales should indicate whether predation is preventing the recovery of the remaining sea lion population.

## Subsistence Harvests

Alaska natives have hunted Steller sea lions since prehistoric times. Archeological sites show that indigenous peoples have harvested sea lions for the past 3,000 to 4,000 years (Laughlin, 1980). Sea lion remains have been confirmed from archeological sites throughout the range of the western stock, including Prince William Sound, Kenai Peninsula, Kodiak Archipelago, and the Aleutian Islands. Steller sea lions comprised up to 70% of the estimated biomass of animal remains at some sites (Haynes and Mishler, 1994). Traditional uses of sea lions include meat for food, hides to cover kayaks, skin of flippers for soles of boots, stomachs for the leg part of boots, intestines for raincoats, and bladders as floats for fishing nets and lines and as sacks to store oil and other liquids (Haynes and Mishler, 1994).

Various traditional hunting methods have been reported, including harpooning on rocks by kayakers in the Kodiak Archipelago (Haynes and Mishler, 1994), gun and spear in the Aleutian Islands (Nelson, 1887), and clubbing on land at Nunivak Island in the Bering Sea (Haynes and Mishler, 1994). Elliott (1887), as cited in Haynes and Mishler (1994), described annual drives that were conducted from September to November on the Pribilof Islands. Hunting was conducted at night when animals were driven into corrals. After a total of 200 to 300 animals were corralled over several nights, sea lions were then herded over 11 miles for 5 days to 3 weeks to the village where they would be slaughtered. Russian orthodox missionaries documented that an average of 2,000 animals were harvested annually on St. George Island in the 1830s (Haynes and Mishler, 1994). By the 1870s, harvests of Steller sea lions from the Pribilof Islands declined to several hundred per year (Nelson, 1887; reproduced in Appendix D).

Reports by subsistence hunters indicate that Steller sea lions declined in abundance in the Aleutian Islands and Bering Sea during the 1800s. In the late 1870s, Pribilof Islanders reported to Nelson (1887) that 70 years earlier sea lions occupied most of the shore of St. George Island and numbered several hundred thousand animals. They further recounted: "By direction of the Russians they were driven off repeatedly until they left the place, and the shore was then occupied by fur seals." By the late 1870s Nelson noted that, while Steller sea lions used to be abundant along the Aleutian Islands, "they are now so scarce among these islands, and the ones that are found there frequent places so difficult to access, that the Aleuts secure very few of them each year." Also, on the Pribilof Islands (referred to at the time as the Fur Seal Islands), Nelson noted that the drive was much more difficult than in previous years, and it was "almost or quite impossible to collect the full number." At the same time, Nelson noted that "they are still rather common at a few points along the north shore of Unimak Island and the peninsula of Alaska" and "eastward and southward they occur all along the coast to California, where their range overlaps that of the southern species [California sea lion]."

During the 20th century, subsistence harvests of Steller sea lions by Alaska natives were documented intermittently. Haynes and Mishler (1994) reported on harvests in selected areas during 1981-1989. The largest documented harvest from one area was 178 animals from six communities on Kodiak Island in 1983. The sporadic coverage of different areas in different years and lack of data from some harvest areas makes it difficult to estimate total statewide harvests by year. The number of Steller sea lions killed by subsistence hunters could be significantly underestimated because not all hunters and communities have been interviewed and there is substantial uncertainty in the estimates of how many animals were lethally shot but not recovered.

The Alaska Department of Fish & Game conducted systematic interviews of approximately 2,100 households from 60 coastal communities to estimate the subsistence harvest of Steller sea lions in Alaska during 1992-1998. Details were reported in annual reports by the department, summarized by Angliss et al. (2001). About 99% of the harvest was taken from the western stock. In 1992 an estimated 549 animals were taken (370 were harvested, 179 were struck and lost). On average, 414 animals were taken (330 harvested, 84 struck and lost) annually during 1993-1995. The reported age composition was 64% males, 19% females, and 17% of unknown sex, and the reported age class composition was 31% adults, 62% juveniles, 3% pups, and 4% unknown. Since 1995 the reported annual takes have declined to as low as 171 animals killed, of which 128 were harvested and 43 were lost.

## DELIBERATE AND INCIDENTAL MORTALITY FROM HUMAN ACTIVITIES

### Shooting of Steller Sea Lions

In the first half of the 20th century, many fox farms existed along the coast of Alaska. Not uncommonly, Steller sea lions were harvested as a cheap source of protein for fox food. During the early to mid-1940s, anecdotal reports revealed incidents of sea lion slaughters by American warplanes stationed in Alaska. In Kodiak "the big PBY's . . . they'd come through the narrows . . . a hundred feet off the water and they'd open up those big fifty caliber guns shooting at the sea lions on Triplet Islands. I was sitting there on the beach . . . and here are these tracers hitting the sea lions and the bluffs with great big sparks flying. The sea lions were dropping off the top of the islands. And man, you talk about slaughter" (Ed Opheim, former cod and salmon fisherman, Kodiak, personal communication, September, 2001).

Shooting of Steller sea lions was legal prior to the Marine Mammal Protection Act of 1972. However, after passage of this act, fishermen were allowed to continue to shoot sea lions that were destroying their gear or causing a threat to human safety. It was not until 1990 when the species was listed as threatened under the Endangered Species Act that the discharge of firearms near Steller sea lions was fully prohibited. Recent court cases and anecdotal information indicate that some illegal shooting of Steller sea lions continues despite these laws.

### Predator Control Programs

In 1951 the Alaska Department of Fisheries instituted a targeted predator control program, mainly for harbor seals but also including sea lions. The rationale was to reduce the impacts of seals and sea lions on the salmon gillnet fisheries. Complete tallies of the number of sea lions killed are unavailable, but unpublished accounts suggest that while a small fraction of the total population was taken, kills on some rookeries were substantial. For example, agency hunters killed virtually all the pups on Amatuli Island in the central Gulf of Alaska during a season once in the 1950s and once in the 1960s.

### Experimental Harvests of Steller Sea Lions

Experimental harvests were conducted from 1959 until 1972 as a more cost-effective alternative to earlier bounty programs for controlling sea lion populations (Thorsteinson et al., 1961). The harvested animals were to be used as food for humans, feed for fox farms in Alaska, and feed for mink ranches in the Pacific Northwest. A total of 45,178 Steller sea lion pups were harvested between 1963 and 1972 from islands in the Kodiak Archipelago to the eastern Aleutian Islands (Merrick et al., 1987). In addition to the experimental harvest, 3,000 to 6,000 pups were taken annually during the 1960s for the fur market (Vania, 1972).

Prior to 1990, shooting was the primary collection method for research on the reproductive biology, physiology, and diet of Steller sea lions. Although a complete count has not been compiled, in several studies biologists reported takes in the range of 80-114 animals (Mathisen, 1959; Calkins et al., 1998).

### Shooting by Fishermen

There is a long history of fishermen shooting sea lions in Alaska (for a review, see Hoover, 1988). In part, shooting was motivated by the belief that declines in salmon runs during the 1930s through the 1950s was partly due to predation by sea lions (Mathisen, 1959). More importantly, fishermen shot many animals in attempts to reduce lost catches and gear damage by sea lions. Sea lions were reported to take fish from a variety of fishing gears, including salmon gillnets, troll and purse seines, herring purse seines, halibut and sablefish longlines, and groundfish trawls (Thompson et al., 1955; Mathisen, 1959; Thorsteinson et al., 1961; Thorsteinson and Lensink, 1962; Hoover, 1988). As early as the 1880s, it was reported that seals and sea lions "prey upon cod, frequently taking them from the line" around Kodiak (Bean, 1887). Sea lions puncture crab pot floats and rip holes in purse seines and gillnets, resulting in lost catch and lost and damaged gear.

At least through the 1970s, lost catch and damaged gear by Steller sea lions were considered to be a serious problem, and there were several attempts to quantify the losses. Based on a poll of halibut fishermen, the International Pacific Halibut Commission estimated that 1.3 million pounds of halibut worth $270,000 were damaged or destroyed by sea lions in 1958 between Cape Saint Elias and the Trinity Islands (Thorsteinson et al., 1961). Matkin (1977) estimated that 8.3% of the salmon catch was damaged during the first 4.5 weeks of the Copper River gillnet fishery in 1977. Three-fourths of the loss was attributed to Steller sea lions, and one-fourth was associated with harbor seals (*Phoca vitulina*). Associated damage to gillnets, nearly all of which was attributed to sea lions, was estimated at 2.5% of the gross value of the catch. During the 1940s to 1970s, there were several published accounts of significant loss of sablefish and halibut to sea lions as longline gear was retrieved (Hoover, 1988).

There is little documentation of the number of Steller sea lions killed by fishermen. A partial survey of salmon trap operators in the Kodiak and Alaska Peninsula areas indicated that 816 animals were killed in spring 1954 (Thompson et al., 1955). Matkin and Fay (1980), as cited by Hoover (1988), estimated that 305 sea lions were shot and killed in association with the drift gillnet fishery at the Copper River delta in 1978. Anecdotal information and public testimony suggest that shooting of animals by fisherman was quite substantial, at least until the 1980s. In the early to mid-20th century, gillnet and other fishermen regularly shot animals that approached their gear, and many people would go to rookeries and haulouts to shoot animals in the Kodiak area (Ed Opheim, former cod and salmon fisherman, Kodiak, personal communication, September, 2001). These shootings appear to have been prevalent throughout coastal Alaska. Later, during the king crab fisheries in the 1960s and 1970s, sea lions were shot for crab bait and crab fishermen shot animals in frustration over punctured floats and lost crab pots. Fishermen active in the joint-venture fishery in Shelikof Strait shot many Steller sea lions to prevent them from taking fish and damaging trawl nets that were towed near the surface for delivery to the motherships. Anecdotal reports by participating fishermen indicate that the number of animals shot was close to the number caught in trawls during those years. Lack of systematic reporting of sea lion kills by fishermen makes it impossible to provide reliable estimates of the impact of shooting on the population (see Box 6.3).

## Incidental Takes and Entanglement in Fishing Gear

Incidental mortality of Steller sea lions in lost (ghost) fishing gear is largely unknown. Most reported entanglements involve fishing net fragments and closed plastic packing bands, such as those used to hold together boxes of bait (Calkins, 1985). Two- to three-year-old animals seem to be most vulnerable. Young sea lions may become entangled in salmon and herring gillnets, but larger individuals are able to break free (Loughlin et al., 1983). Loughlin et al. (1986) examined approximately

## BOX 6.3
## Anecdotal Mortality

There is a long history in Alaska of shooting sea lions for their meat and pelts, for sport, or to reduce sea lion interference with fishing operations. Prior to 1990, a prevalent attitude among many fishermen was that sea lions were a nuisance and a cause of damaged gear and lost catches. Currently, the presence of observers on factory trawlers makes it unlikely that these fisheries illegally shoot sea lions. The number of sea lions killed in other fisheries is unknown, but even small numbers of shootings could contribute to the decline of the small remaining populations. Except for subsistence harvest, shooting of sea lions has been illegal since 1990. Still, the committee heard many anecdotal accounts of unreported "takes" of Steller sea lions, suggesting that the tradition may continue at a higher incidence than the recent estimates (Loughlin and York, 2000). The following examples are from a variety of unconventional sources. Shooting, both legal and illegal, will have a particularly severe impact on the dwindling population, and better estimates of this mortality are urgently required.

- A captain on a U.S. Fish and Wildlife Service research vessel recounted that when he first arrived in Alaska in the mid-1970s he crewed on a crab fishing vessel in the Bering Sea. On the first encounter with Steller sea lions, the skipper pulled out a rifle and started shooting. Apparently this behavior was commonplace—crab fishermen loathed sea lions because the sea lions often punctured their buoys, causing them to lose the gear.
- A wildlife guide on St. Paul Island in the Pribilofs was told that local inhabitants would often shoot swimming sea lions from shore in the hopes that dead sea lions would wash up on the beach. Some local residents estimated that only 1 out of every 50 animals were recovered (2%). However, an average recovery rate of 80% (1993-1995) was reported by the Alaska Department of Fish & Game based on interviews of coastal community residents, potentially underestimating the full impact of the subsistence harvest.
- Recently, a Russian orthodox priest who worked for years on Kodiak Island thought there was no mystery at all to the continuing decline of Steller sea lions. During confession, he often heard fishermen talk about shooting sea lions.
- The Alaska SeaLife Center in Seward had planned a public display of Steller sea lion, to be viewed from a plaza. The tank was initially designed so the public could view the animals from outside the building. However, the plan was abandoned out of concern that the sea lions would be shot if they were visible from the street.
- A fisher from southeastern Alaska claimed she currently knew at least 25 fishermen who still shoot virtually all Steller sea lions on sight. Although she knew that the eastern stock seems to be healthy and increasing, she wondered if such activities could explain the decline of the western stock.
- In May 1993 a university student researcher saw eight recently shot Steller sea lions in the nearshore area of a bay in the eastern Bering Sea. Two hunters were convicted in the case.
- A committee member was sport fishing in southeastern Alaska in July 2002 and met a fisherman who stated that he had excellent salmon fishing but that most of his fish were being taken by Steller sea lions. He was disgruntled that he did not have "seal bombs" aboard and had to move to another area with poorer catch rates.

16,000 adults and 14,000 pups at 17 rookery and 15 haulout sites in the central and eastern Aleutian Islands in summer 1985. Only 11 animals showed evidence of entanglement; most involved nets or twine. However, these were individuals that survived the encounter, and mortality due to entanglement is unknown. The amount of netting and packing straps varies with beach location and exposure on Kodiak Island (Calkins, 1985), and entanglement rate might be expected to vary with type and amount of fishing gears, which vary seasonally by area. Entanglements of Steller sea lions on longline hooks (Hoover, 1988) and salmon troll gear (Angliss et al., 2001) are occasionally reported.

Steller sea lions may become captured during commercial and sport fishing operations. Most accounts of sea lion deaths were reported during the foreign trawl fisheries, which undoubtedly have caught Steller sea lions since their inception (Perez and Loughlin, 1991). Unfortunately, recorded observational data are limited to onboard observer programs restricted to certain fisheries, vessel size classes, and years of operation. Hence, recorded takes underestimate the total number of sea lions killed in the commercial fisheries. Observers aboard foreign and joint-venture trawl vessels recorded 3,661 marine mammals (90% were Steller sea lions) as incidental takes during 1973-1988 (Perez and Loughlin, 1991). The highest catches of sea lions were taken in a joint-venture pollock trawl fishery in Shelikof Strait in 1982-1984 and foreign fisheries near Kodiak and in the Aleutian Islands in earlier years. Also, sea lions were commonly taken in foreign trawl fisheries in the southeastern Bering Sea in 1978-1981 (Loughlin et al., 1983). In Bering Sea fisheries, most incidentally caught Steller sea lions were males, whereas in the Gulf of Alaska and probably in the Aleutian Islands most takes were females and subadult males (Loughlin and Nelson, 1986; Perez and Loughlin, 1991).

To estimate total incidental mortality, observed sea lion takes must be adjusted for sea lion survival rate, fraction of observer days spent recording sea lion takes, and fraction of the fleet with onboard observers. This assumes that sea lion takes were the same on unobserved vessels, an assumption that could underestimate the number of takes if fishing practices are different in the absence of an observer. In 1979, 34% of incidentally caught sea lions were alive when brought onboard foreign trawlers and escaped into the water when released from the net (Loughlin and Nelson, 1986). In contrast, sea lion mortality on U.S. trawlers participating in the joint-venture fishery was nearly 100% because net codends were tied off after retrieval and remained in the water a long time before transfer to the foreign processing vessel.

Annually, about 800 marine mammal mortalities (predominantly sea lions) were estimated for foreign fisheries in Alaska during 1972-1976. This most likely underestimates mortality because there were a limited number of observers during this time period and reporting protocols were informal and variable (Loughlin et al., 1983). During 1978-1981, only 10% of foreign fishing vessels (mean = 251 vessels per year) carried observers. The estimate of 800 annual mortalities does not include animals taken by U.S. domestic fishermen, including those who participated in the joint-venture fisheries. During the 1980s, observer coverage rates approached 100% for foreign trawl and longline fishing vessels and foreign processing vessels participating in joint-venture fisheries with U.S. catcher boats.

In 1980 a joint-venture fishery developed to harvest large schools of spawning pollock in Shelikof Strait (Loughlin and DeLong, 1983; Loughlin and Nelson, 1986). In this fishery alone, total mortalities of sea lions were estimated to be 958 to 1,436 in 1982, 216 to 324 in 1983, and 237 to 355 in 1984 (Loughlin and Nelson, 1986). During these years, a single tow was observed with more than 20 sea lions, but occasionally fishermen reported takes of 50 to 100 sea lions in a single tow.

Particular aspects of the joint-venture pollock fishery contributed to high incidental mortalities of Steller sea lions in Shelikof Strait in the early 1980s. Roe was the most valuable seafood product, and large volumes of fish were discarded, perhaps attracting sea lions to the area (Loughlin and DeLong, 1983). Nets were retrieved, sometimes with many sea lions in the vicinity, and the method of transfer of codends from catcher to processing vessels led to the drowning of ensnared sea lions.

A number of factors contribute to the variability in rates of incidental takes of sea lions by the joint-venture trawl fishery for pollock in Shelikof Strait in the 1980s (Loughlin and Nelson, 1986). Fishing locations varied substantially; compared to 1983 and 1984, the fishery in 1982 occurred farther east and closer to shore, where presumably more sea lions were encountered (see Figure 6.3). Also, more sea lions were observed in Shelikof Strait in mid- to late April than in January to March. During 1982-1984, 80% of caught sea lions were taken in April. Presumably, sea lions move into the area in greater numbers in April in advance of the pupping and breeding season, which begins in May. The highest catch rates of sea lions in 1982 may be partly explained by the persistence of this fishery in April, compared to the termination of the fishery in early April in 1983 and 1984.

Time of day is another important factor affecting the bycatch of Steller sea lions. In the joint-venture fishery, trawl tows were made at all hours of the day, but only 18% of the catch of Steller sea lions occurred during daylight (see Figure 6.4). Most sea lions were caught during a 4-hour period between 11 p.m. and 3 a.m.

During the 1990s, incidental takes of Steller sea lions were monitored for domestic groundfish trawl, pot, and longline fisheries conducted in the Gulf of Alaska, Aleutian Islands, and Bering Sea (Angliss et al., 2001).

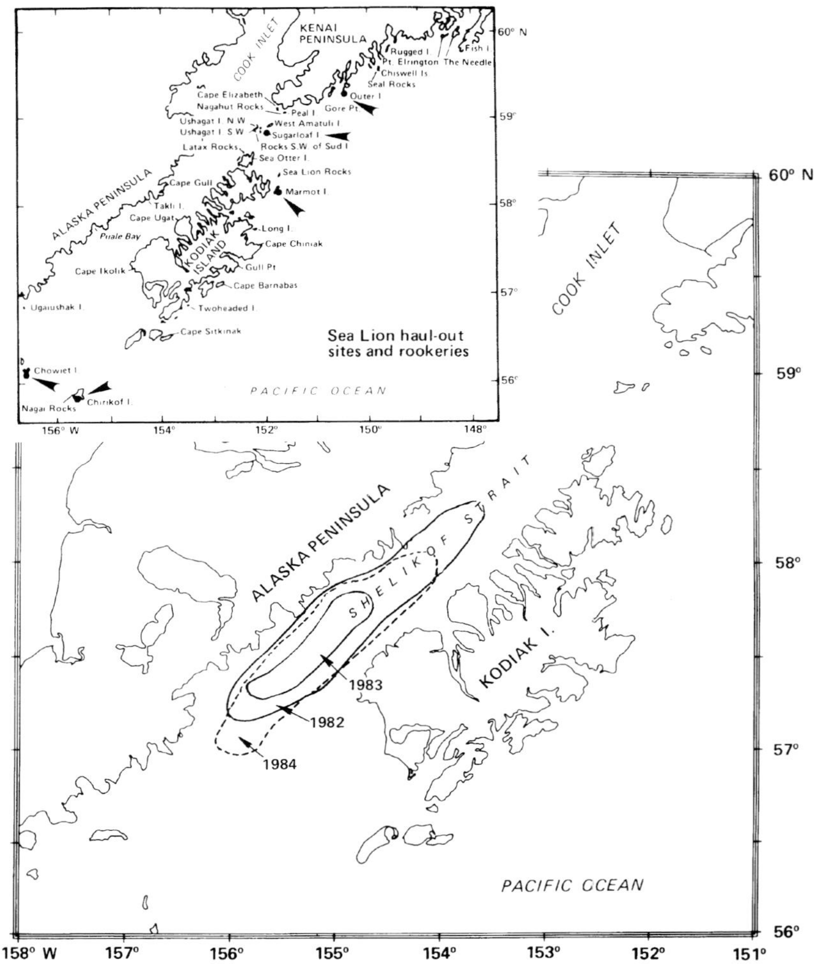


FIGURE 6.3 Generalized areas of incidental catch of Steller sea lions during the pollock joint-venture trawl fishery in Shelikof Strait in 1982-1984. Locations on the inset depict sea lion rookeries (arrows) and haulouts.
SOURCE: Loughlin and Nelson (1986); reprinted with permission from the Society of Marine Mammalogy.

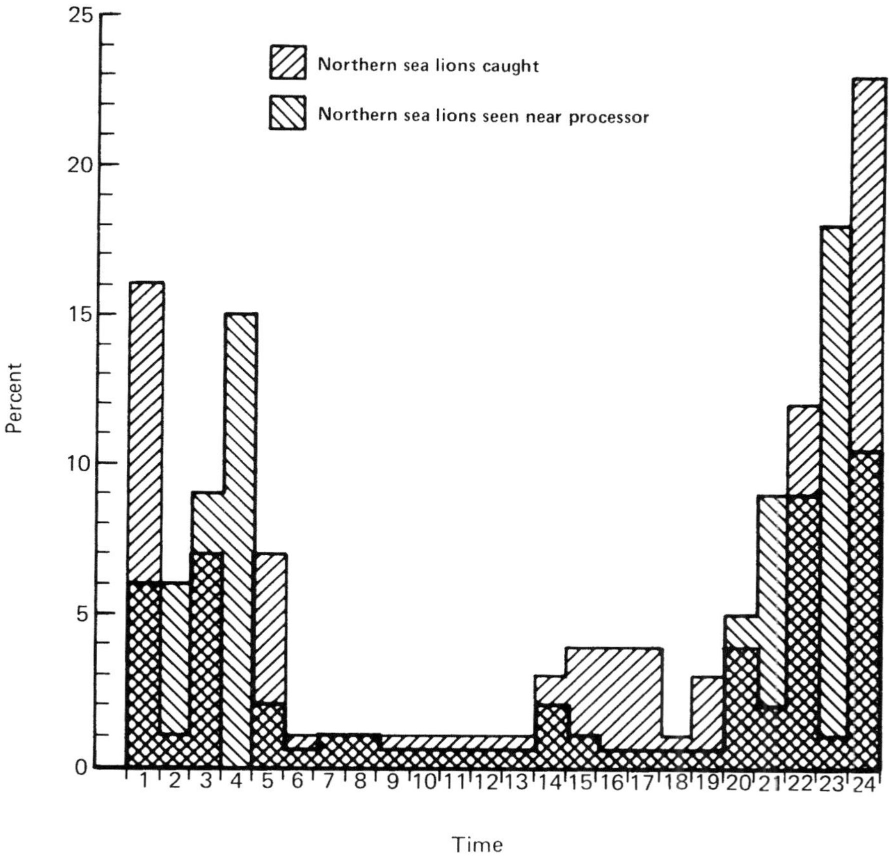


FIGURE 6.4 Percent of incidental catches of Steller (Northern) sea lions by time of day (hour) compared to percent of Steller sea lions seen near processing vessels, as reported by observers in the joint-venture pollock fishery in Shelikof Strait in 1983-1984.
SOURCE: Loughlin and Nelson (1986); reprinted with permission from the Society of Marine Mammalogy.

No mortality has been observed in the pot fisheries since 1990 and in the longline fisheries since 1995. The one observed mortality in the longline fishery was expanded by the fraction of the observed fleet to estimate a total mortality of four animals. The highest mortality from any gear/region combination was from the Bering Sea trawl fishery, where it was estimated that seven sea lions were killed annually during the 1990s.

A salmon drift gillnet fishery with 4 to 5% observer coverage in 1990-1991 recorded two sea lion deaths. Approximately 2 to 5% of the salmon setnet and gillnet fisheries were observed in Cook Inlet in 1999 with no reported mortalities of sea lions.

In summary, low rates of incidental fishing mortality of Steller sea lions have been reported for the 1990s. However, observers do not cover all fisheries, and the only information on unobserved fisheries comes from self-reporting of marine mammal takes or beach standings of entangled animals. All such reports are voluntary.

### Summary of Effects of Kills by Humans

It is not clear whether apparent declines in Steller sea lions in the Aleutian and Pribilof Islands between the 1830s and 1870s were related to subsistence harvests or other factors. For the next 100 years, good data on subsistence harvest are lacking. Contemporary data on subsistence harvests are available only for the 1990s, and these harvests represent less than 1% of the abundance of the western stock of Steller sea lions. However, total mortality from subsistence harvest could be higher if the number of unrecovered animals is higher than the current estimates. At this time, the subsistence harvest is not considered to be a major contributing factor to the current decline (National Marine Fisheries Service, 2000).

It is difficult to fully assess the effects of incidental mortality from fisheries because there were low rates of observer coverage during the 1970s and early 1980s. Using all available observer data, Perez and Loughlin (1991) estimated that annual incidental mortality of Steller sea lions might have declined from approximately 1,000 to 2,000 in the early 1970s and 1982 to fewer than 100 animals in 1988. If these estimates are representative of total mortality from all fisheries, then fishery-induced mortality explains only a minor portion of the decline of the western Steller sea lion population. However, extrapolations of total kill rates depend heavily on assumptions about unobserved foreign and domestic vessels and fisheries during these years. The true number of kills is unknown but could be considerably higher than official estimates. Also, although poorly documented, it appears indisputable that historical shooting of sea lions by fishermen and others was commonplace and likely represented a very large number of animals. Because it has been illegal to discharge firearms within 100 yards of Steller sea lions since 1990, the amount of shooting in recent years is even less well known, but some anecdotal evidence suggests that the practice persists. Hence, the "smoking gun" in the mystery of the Steller sea lion population decline could, at least in part, be the unreported and illegal takes.

## INFECTIOUS DISEASE

Theoretical studies suggest infectious diseases may regulate host abundance by exerting density-dependent effects on host reproduction or survival (Anderson, 1979). However, few empirical studies on free-living animals have determined whether effects of disease on populations are density dependent, as most disease investigations on wildlife have focused on determining the proximate causes of large die-offs. Such die-offs are most often the result of epidemics of disease in host populations that have not previously been exposed to the disease or that have become more susceptible to an introduced agent due to changes in immune status. Although not generally considered regulators of host population density, severe epidemics may reduce host population density to such an extent that stochastic events or previously unimportant ecological factors may further reduce the host population size (Harwood and Hall, 1990).

The importance of infectious disease epidemics in causing declines of marine mammal populations is unclear because few die-offs have been investigated sufficiently to determine their cause, and it is often difficult to accurately determine host population numbers. Recent epidemics in marine mammals have caused dramatic mortality, but the effects on host population numbers vary. For example, approximately 17,000 harbor seals (70% of the population) died in the phocine distemper (PDV) epidemic in Europe in 1988, but 10 years later the population had recovered to pre-epidemic numbers (Reijnders et al., 1997). The source of infection for this morbillivirus outbreak in marine mammals is unclear. The 1988 PDV epidemic was believed to have resulted from the introduction of a virus into a naïve population, due to the large numbers of animals without antibodies prior to the outbreak. Based on this assumption, a mathematical model investigating the infection dynamics of this disease in 1992 predicted that reintroduction of the virus resulting in large-scale mortality would not occur for at least 10 years (Grenfell et al., 1992). A new outbreak of PDV is currently occurring in the North Sea (Jensen et al., 2002; *cwss.www.de/news/news/Seals/01-seal-news.html*). In 1988 it appeared the mortality was lower in seal populations that had experienced previous mortality events (Grenfell et al., 1992).

Endemic diseases have more subtle effects and may be more important in regulating marine mammal populations than previously thought.

Infectious diseases such as brucellosis are known to cause population declines in terrestrial mammals due to spontaneous abortion and reproductive failure. Although these types of disease organisms have only recently been isolated from marine mammals, they may be prevalent in free-ranging populations (Dunn et al., 2001). Food limitation may increase the impact of macroparasites, resulting in population crashes, as observed in Soay sheep populations in Scotland (Gulland, 1992). Even if only a few individuals show signs of infection, it is possible that parasitic nematodes have a significant influence on the host population size (Hudson and Dobson, 1995). There are a wide variety of parasitic nematodes in marine mammals, but their effects on host population dynamics are unknown (Dailey, 2001).

A number of infectious disease agents known to occur in marine mammals of the Arctic and Pacific could cause epidemics, endemic mortality, or decreased reproduction in Steller sea lions. These include morbilliviruses, influenza virus, phocine herpesviruses, caliciviruses, *Leptospira* spp., *Brucella* spp., *Chlamydia psittaci*, *Toxoplasma* spp., and various species of nematodes (Dierauf and Gulland, 2001). If an epidemic hit a population of Steller sea lions, it would be expected to spread from one area, cause mortality in animals of all ages, and leave survivors with antibodies to the causative agent (Heesterbeek and Roberts, 1995). Although the rapid decline of sea lions is consistent with the first two of these conditions, to date, no antibodies have been detected to morbilliviruses or influenza virus—the two viruses most likely to cause such an epidemic (Burek et al., 2001). Antibodies to these viruses are also absent from sea otters (*Enhydra lutris*) and northern fur seals (*Callorhinus ursinus*) that share waters around the Aleutians with Steller sea lions and are susceptible to these diseases (Hanni et al., in press; Terry Spraker, Colorado State Veterinary School, Fort Collins, personal communication, 2000). Thus, it is unlikely, but not impossible, that a viral epidemic caused the rapid decline of Steller sea lions during the 1980s. If morbillivirius did cause an undetected epidemic, however, experience from the European harbor seal epidemics suggests that it would not continue to affect population size in the absence of other factors and that the population should show signs of recovering to pre-epidemic numbers. To definitively eliminate the hypothesis that a viral epidemic caused the rapid decline of Steller sea lions would require more comprehensive tests of banked serum samples and molecular tests for the presence of disease agents in preserved tissue samples.

Endemic diseases could inhibit recovery of the Steller sea lion population. Antibodies to phocid herpesvirus, caliciviruses, and leptospires have been detected in Steller sea lions (Barlough et al., 1987; Calkins and Goodwin, 1988; Zarnke et al., 1997; Burek et al., 2001). All of these agents cause mortality or reproductive failure in other marine mammal species, but their effects on Steller sea lions have not been documented because neither sick nor dead animals have been available for examination. In other marine mammals, these endemic diseases cause reproductive failure through abortions or poor pup survival. Higher levels of haptoglobin, a nonspecific indicator of inflammation, have been detected in Steller sea lions in declining populations (Zenteno-Savin et al., 1997). Inflammation is often a consequence of infectious disease; therefore, higher haptoglobin levels could be an indication that declining populations of sea lions have higher rates of infection. The prevalence of these disease organisms in the Steller sea lion population is unknown. Although antibodies to some disease agents have been found in sera, few studies have confirmed the presence of the disease-causing agent in tissue samples. *Chlamydia* spp. was observed in tissues of an aborted fetus (Spraker and Bradley, 1996), and antibodies to this organism are widespread in adult Steller sea lions, especially females (Burek et al., 2001). Further studies should be directed at determining the prevalence of infection in Steller sea lion populations and the effect of infection on reproduction and survival.

Macroparasites such as nematodes, flukes, and tapeworms are common in pinnipeds and may cause mortality in malnourished animals. California sea lions (*Zalophus californianus*) that suffer food deprivation or feed on unusual prey species during El Niños have heavier parasite burdens than animals in other years and can die from parasitic ulcers (Fletcher et al., 1998). These macroparasites have been found in Steller sea lions, possibly compounding the effects of malnutrition and increasing juvenile mortality. Little is known about the species of parasites or their prevalence and intensity of infection in Steller sea lions. Future studies should identify macroparasites, determine the prevalence and intensity of infection, and determine whether infection intensity correlates with nutritional status.

In conclusion, little is known about the prevalence of infectious diseases in Steller sea lions or their morbidity. Both eastern and western populations of Steller sea lions have antibodies to agents that could decrease survival and reproduction. The prevalence and intensity of infections need to be assessed to determine whether they play a role in the decline of Steller sea lion populations. Although a viral disease could have occurred in the 1980s, to date there is no direct evidence of an epidemic.

## TOXINS

Biotoxins produced by harmful algal blooms have caused episodic mortality in a number of marine mammal populations around the United States, from manatees (*Trichecus manatus*) off Florida to California sea lions off California (Bossart et al., 1998; Scholin et al., 2000). A bloom of saxitoxin-producing algae is believed to have caused a die-off of about 50% of the Mediterranean monk seal population, although controversy still surrounds this event due to the lack of fresh carcasses for examination (Hernández et al., 1998). Harmful algal blooms producing toxins such as domoic acid occur in Steller sea lion foraging habitats, but their effect on these animals is unknown because no carcasses have been found or examined. It is unlikely that a large mortality event occurred, however, because these toxins cause dramatic clinical signs that would have been readily detected. In addition, mortality of other species (including fish and birds) usually occurs, and these were not reported in areas of sea lion declines. Furthermore, to account for the pattern of sea lion decline, the bloom of toxic algae would have to spread from the central Gulf of Alaska to the western Aleutians but not to southeastern Alaska. It is unlikely that such an event would have gone unnoticed. However, retrospective analysis of stored tissues for biotoxins would be necessary to completely rule out the possibility that algal toxins contributed to the rapid decline of Steller sea lions.

There is an extensive literature on the effects of toxic contaminants on mammalian reproduction (reviewed by O'Hara and O'Shea, 2001). There are also data on the levels of a number of elements in marine mammal tissues (e.g., cesium, cadmium, mercury, selenium) and persistent organic compounds such as polychlorinated biphenyls (PCBs) and dichloro-diphenyl-trichloro-ethane (DDT), but few data exist on dose-response effects even for well-known contaminants in marine mammals. It is thus not possible to determine whether the levels of contaminants measured in tissue samples affect the survival of Steller sea lions. The levels of some xenobiotic compounds have been determined in a limited number of Steller sea lion tissues. PCB and DDT levels in blubber of sea lions sampled between 1976 and 1981 in the Bering Sea were lower than in sea lions from the Gulf of Alaska (Lee et al., 1996). Levels were lower in females than males, as occurs in most marine mammal species, due to the lactational transfer of lipophilic toxins. Levels of PCBs and DDTs were higher in Steller sea lions than in ringed and harp seals from Arctic waters but were comparable to levels in gray seals from the east coast of Canada and lower than in California sea lions with normal gestation periods. Both gray seals and California sea lion populations are currently increasing. It is thus unlikely that the contaminant levels in Steller sea lions are causing direct mortality in this species, although more subtle effects on physiology could occur. There may be species-specific effects, and combinations of contaminants may have more deleterious effects than single compounds. Thus, it is not possible to eliminate the possibility that contaminants affect the physiology of Steller sea lions by measuring a few compounds in blubber at any one time during development.

Positive associations between organochlorine burdens and reduced immune function have been observed in harbor seals, but the overall effect of the health of the population is still unclear (deSwart et al., 1994). If contaminants were causing immunosuppression in Steller sea lions, an increase in prevalence and susceptibility to infectious disease should be observed in declining populations, but these epidemiological observations are lacking. Differences in contaminant burden have been inferred from fecal levels of PCBs in Steller sea lions that could result from regional differences in the prey population (Beckmen et al., 2001). Because there has been considerable military activity in the Aleutians, it is possible that certain sites have localized contamination with unidentified compounds. Estimates of vital rates of Steller sea lions in different locations may uncover differences in local mortality and reproduction that are indicative of toxic contamination. Further epidemiological studies focusing on associations between contaminant levels in tissues of individuals and life history parameters, coupled with determination of the significance of reproductive failure and infectious disease in the dynamics of Steller sea lions are needed to determine whether contaminants could play a role in limiting sea lion recovery.

## WEIGHT OF EVIDENCE

Steller sea lion behavior, physiology, life history, and environment can be analyzed with regard to how they would be expected to change under each of the eight hypotheses described above using a simple positive or negative response variable (Bowen et al., 2001). In the analysis presented below, the disease category is divided into epidemics and endemics because these two processes are expected to produce different responses in sea lions. Biotoxins are included in the epidemic disease category because the response variables should change in the same direction.

As discussed in the Bowen et al. review, the directional changes represent the most likely responses based on current information but do not represent predictions. Use of this analysis should be tempered by consideration of the following:

1. For some response variables the direction of change under a given hypothesis is debatable. Change may depend on whether the effect is size selective, is local or regional, or reduces performance rather than increases mortality (e.g., disease, pollution).

2. The magnitude of change may vary with the intensity of the effect. Hence, it is possible to have a false correlation if an effect appears to be consistent but is too minor to affect population size.

3. Data used to determine the direction of each response variable may be biased due to sampling errors such as age or size-specific effects, local effects generalized to the whole population range, or time-frame dependence (what is observed in one year becomes generalized over a decade).

In analyzing the weight of evidence for each hypothesis, the committee assumes that each variable changes as the result of direct interaction. The various hypotheses listed in Table 6.1 can be characterized as either bottom-up or top-down forcing mechanisms (e.g., Hunter and Price, 1992). Bottom-up forcing hypotheses are defined by their impact on sea lion food availability and include biomass removed by fisheries, climate change, and regime shifts. Top-down forcing hypotheses are defined by their impact on survival (assuming food is not limiting) and include disease, predation, and human killing of all kinds. Pollution has impacts that could reflect either top-down or bottom-up effects. Grouping hypotheses simplifies the task of pattern assessment because the response variables are predicted to change in a consistent manner depending on whether the direction of forcing is bottom up or top down. For example, foraging time should increase when per capita food availability decreases because prey has been depleted by fishery removals or productivity has changed due to a climate regime shift. Foraging time should decrease as per capita food availability increases if predation or human takes reduce the size of the sea lion population relative to the prey base.

The committee extended this analysis by including the available observational evidence for comparison with the expected direction of change (Table 6.2). The directionality of the observed response is determined by comparison of sea lions in the western population with sea lions in the eastern population. For this purpose, characteristics of sea lions in the eastern population are assumed to be representative of the western population prior to the start of the decline. The rationales for the expected response and data sources for the observed response are described in Box 6.4

Table 6.2 lists relevant behavioral, physiological, and life history metrics of Steller sea lions, and features of the associated ecosystem for which data are available. The expected directions of change were derived from observations or ecological models as first described in the "Is It Food?" workshop (Alaska Sea Grant College Program, 1993). Eberhardt (1977) used similar reasoning to provide a general framework for assess-

TABLE 6.2. Matrix of Expected/Observed Directional Changes in the Response Variables Under Hypotheses Proposed to Explain the Decline of the Western Steller Sea Lion Stock. The response variable changes are given as either higher (H) or lower (L), for example under a bottom-up forcing, the predicted impact of fishery removals is lower (L) birth mass. However, higher (H) birth masses have been observed. Therefore the entry in the table is L/H. Matches between observed and expected (excluding unknowns) are in bold.

| Forcing Direction | Bottom-up | | Top-down | | | | | Uncertain |
|---|---|---|---|---|---|---|---|---|
| Correlate/Response Variable | FR | CE | PRED | DT | SH | IT/ENT | D | PO |
| *Pups* | | | | | | | | |
| Birth mass (1) | L/H* | L/H* | **H*/H*** | **H*/H*** | **H*/H*** | **H*/H*** | U/H* | U/H* |
| Pup growth rate (1,2[a],3[a]) | L/H | L/H | **H*/H** | **H*/H** | **H*/H** | **H*/H** | U/H | U/H |
| *Adult female* | | | | | | | | |
| Body condition (4,5,6) | L/H | L/H | **H*/H*** | **H*/H** | **H*/H** | **H*/H** | L/H | L/H |
| Foraging trip duration (7,8) | H/L | H/L | **L/L** | **L*/L** | **L*/L** | **L*/L** | **L/L** | **L/L** |
| Dive depth (9) | H/L | H/L | **L/L** | **L*/L** | **L*/L** | **L*/L** | **L/L** | **L/L** |
| Field metabolic rate (9) | H/L | H/L | **L*/L** | **L*/L** | **L*/L** | **L*/L** | **L/L** | **L/L** |
| *General* | | | | | | | | |
| Foraging range (10) | H/L | H/L | **L/L** | **L/L** | **L/L** | **L/L** | **L/L** | **L/L** |
| Beach strandings (11) | H/L | H/L | **L/L** | U/L | **L/L** | **L/L** | H/L | H/L |
| Other piscivores (12) | L/NC | L/NC | **NC/NC** | **NC/NC** | **NC/NC** | **NC/NC** | **NC/NC** | U/NC |
| Food availability (10) | L/H | L/H | **H/H** | **H/H** | **H/H** | **H/H** | **H/H** | **H/H** |

FR—fishery removals
PO—pollution
CE—climate/regime shift
PRED—predation
DT—direct take (shooting)
SH—subsistence harvest
IT/ENT—incidental take/entanglement
D—disease
H—higher
H*—higher or no change
L—lower
L*—lower or no change
NC—no change
U—uncertain

[a]Based on pup mass at one month.

SOURCES: Data came from (1) Brandon and Davis (1999); (2) Merrick et al. (1995); (3) Rea et al. (1998); (4) Davis et al. (1996); (5) Adams (2000); (6) Mike Castellini, University of Alaska, Fairbanks, personal communication, 2002; (7) Brandon (2000); (8) Milette (1999); (9) Andrews et al. (2002); (10) National Marine Fisheries Service (2001); (11) Calkins et al. (1998, p. 241); and (12) Dragoo et al. (2000). Table was modified from Bowen et al. (2001).

## BOX 6.4
## A Brief Rationale for the Directions of Population Change in Table 6.2

***Birth mass***—Reduced food availability is expected to limit a female's ability to expend energy on her offspring. Bottom-up forcing mechanisms should thus lead to a reduced birth mass, while top-down forcing mechanisms, if per capita food availability is not limiting, should lead to increased birth mass. The directional effects of disease and environmental pollutants are indeterminate: if these factors increase morbidity, the outcome (fewer sea lions and less competition) would resemble a top-down mechanism, but if they reduce fitness, the outcome (less energy available during gestation), would resemble a bottom-up mechanism. Comparisons of pup birth mass in declining rookeries with birth weights in stable or increasing rookeries in southeastern Alaska during the 1990s have shown that birth masses are similar (Brandon and Davis, 1999).

***Pup growth rate***—An argument similar to that made for expected changes in birth mass. Pup growth rates (Brandon and Davis, 1999) and masses at 1 month (Merrick et al., 1995; Rea et al., 1998) were higher in the western population.

***Body condition***—An argument similar to that made for expected changes in birth mass, except that disease and environmental contaminants should lead to reduced body condition. Adult females in the west have been observed to have greater mass and more body fat (Davis et al, 1996; Adams, 2000; Michael Castellini, University of Alaska, Fairbanks, personal communication, 2002).

***Foraging trip duration***—As per capita food availability declines, the time needed to obtain nutritional resources necessary for maintenance and reproduction should increase. Therefore, bottom-up forcing scenarios suggest increased foraging trip duration, while top-down forcing mechanisms, by increasing per capita food availability, suggest reduced foraging trip duration. Foraging trip length for females with pups was lower at western rookeries in the mid-1990s (Brandon and Davis, 1999; Andrews et al., 2002).

***Dive depth***—Diving is costly in terms of time and energy expenditure and, in the case of deep dives, is the less efficient metabolic pathway associated with anaerobic diving. Thus, sea lions should dive no deeper than necessary to obtain the nutritional resources needed for maintenance and reproduction. Increased dive depths are consistent with bottom-up forcing scenarios, and reduced dive depths are consistent with top-down forcing mechanisms, both through effects on per capita food availability.

***Foraging range***—An argument similar to that made for expected changes in foraging trip duration and dive depth. Measurements of foraging range in 1997 at a rookery in the central Aleutians and one in southeastern Alaska indicate that foraging range is lower in the west (Andrews et al., 2002).

***Field metabolic rate***—As per capita food availability declines, the search and pursuit times required to obtain nutritional resources needed for maintenance and reproduction should increase—hence, an argument similar to that made for expected changes in foraging behavior. Andrews et al. (2002) measured metabolic rate at the same rookeries as described above and found a lower rate in the Aleutians.

***Beach strandings***—Expected changes in the number of stranded or moribund sea lions vary less consistently across categories of forcing mechanisms and in some cases are contingent on a variety of important details. Most forms of food limitation should lead to increased numbers of weakened and emaciated animals onshore. Predation by killer whales (assuming that attacks are rarely unsuccessful and that all or most of the carcasses are consumed) produce no or minimal visual evidence. In contrast, shark predation is associated with diagnostic scars on pinnipeds in areas where shark attacks have been observed. If shooting occurs on- or near shore, elevated numbers of stranded carcasses would be expected, even though sea lions normally sink after being shot and killed. Shooting farther from shore might result in the carcasses sinking or decomposing before they drift ashore. These same arguments apply to subsistence harvest and incidental losses in fishing gear. Eventual mortality from wounds inflicted during a subsistence hunt might lead to elevated numbers of beached carcasses, whereas animals dying from entanglement in fishing gear would likely be too far from shore for this to happen.

***Other piscivores*** *(surface feeding and diving piscivorous seabirds)*—Since piscivorous seabirds feed on earlier life stages of many of the same prey as Steller sea lions, their populations also would be expected to decline under a bottom-up forcing scenario. Most top-down forcing scenarios, in contrast, predict no change in seabird populations. Killer whales or sharks in the northern hemisphere do not consume seabirds, at least so far as is known. A subsistence harvest for pinnipeds should have little effect on seabird populations. The effects of gear entanglement and incidental takes in fisheries is less certain due to the substantial differences in body size between seabirds and marine mammals. Nonetheless, the most likely direct impacts of significant incidental mortality on seabirds in fishing gear are reduced populations. Disease in Steller sea lions would be expected to leave seabird populations unchanged. The expected effects of environmental contaminants are less certain—direct effects on sea lions should leave seabird populations unchanged, whereas effects on the prey base might cause both to decline.

***Food availability***—Prey abundance should decline under the bottom-up forcing scenarios and increase under the top-down scenarios. The effect of environmental contaminants on food availability is unknown.

ing marine mammal population status. Box 6.4 provides a brief rationale for the directions of change given in Table 6.2.

## SYNOPSIS

If the rationale for the expected patterns of decline (Box 6.4) is generally correct, the weight of evidence for causality of the Steller sea lion decline since 1990 is most consistent with a top-down forcing scenario. Predation and human-induced mortality provide a good fit to the available data. Currently available assay data for disease and contaminants do not indicate additional mortality from these factors. The virtual absence of beach strandings at first appears inconsistent with unreported shooting. However, it is known that most Steller sea lions sink after being shot in the water. In their study of sea lion stomach contents, Imler and Sarber (1947) shot at least 20 animals in the water and only one (<5%) floated. Likewise, in another feeding study, Fiscus and Baines (1966) killed 34 animals in the water and 23 (68%) sank before they could be recovered. It is possible that the continuing decline in the population is caused by a combination of mortalities from killer whale predation, illegal shooting, incidental takes in fishing gear, and subsistence harvest.

Evidence gathered in the 1990s is generally incompatible with a bottom-up forcing scenario. Indicators of sea lion health and foraging behavior suggest that the western population is not food limited when compared to the stable, slowly increasing population in southeastern Alaska. Indirect effects of fishing or other ecosystem changes may influence sea lion population trends (Box 6.1), but support for these potential mechanisms would require a more in-depth understanding of food web interactions in the region.

During the rapid decline of the sea lion population in the 1980s, some studies suggested a decrease in sea lion fitness consistent with a bottom-up forcing scenario. Although groundfish fishery yields were high during this period, the biomass of these species was also high. Hence there was no global reduction in the amount of groundfish available to sea lions. The possibility remains that local depletion of some fish stocks, such as Atka mackerel, may have occurred in Steller sea lion habitat, but there is less support for local depletion of pollock stocks. Changes in the abundance of forage fish species related to the regime shift in the late 1970s may have affected sea lion fitness, but these effects do not appear sufficient to account for the large mortality of sea lions in the 1980s. Although a disease epidemic could explain a rapid drop in population, to date there is no indication of such an event based on immunological analysis of serum samples. Similarly, available data do not support widespread mortality from biotoxins or contaminants. Top-down sources of mortality also apply to the earlier phase of the decline, and in the case of human takes, the levels are known to be higher before 1990 than in the most recent decade. Shooting of sea lions was legal to protect fishing gear, and high levels of bycatch mortality were reported in the joint-venture fisheries during the 1980s. Also, it is possible that predation by killer whales was higher than previously estimated. Therefore, even during the rapid decline of the western population, it is likely that a combination of top-down and bottom-up forcing mechanisms were responsible for the high mortality of Steller sea lions.

# 7

# Information Needs and Recommendations

Two paradoxes permeated the committee's discussion. First, the available behavioral and physiological evidence for the 1990s suggests that individuals in the declining western stock of Steller sea lions are at least as healthy as individuals in the eastern stock, which has been increasing by about 2% per year. Second, the western stock has not shown signs of recovery despite substantial restrictions imposed on commercial groundfish fisheries since 1999. This has raised questions about the presumed interaction between sea lions and commercial fishing activity. The available data are inadequate either for fully justifying or for overturning fishery management measures to protect Steller sea lions. Therefore, the National Marine Fisheries Service (NMFS) and the North Pacific Fishery Management Council still will be obligated to make regulations based on inconclusive information. Because many of the disputes over management actions (the reasonable and prudent alternatives) arise from the divergent interests of stakeholders, resolution may not occur until there is considerable evidence confirming one (or more) of the hypotheses or there is a clear and sustained recovery of the Steller sea lion population.

## CURRENT AND FUTURE INFORMATION NEEDS

The ability to determine the effects of restrictions on fisheries on recovery of the Steller sea lion population will depend not only on the biophysical data that are collected but also on the management regime that is in place when the data are collected. If present closures are maintained or expanded in response to demands for a more precautionary approach to Steller sea lion protection, it will likely be difficult or impossible to attribute future population changes to the new management regime. This difficulty arises because these management changes will take place in a continuously varying physical and biological environment. To complicate matters further, Steller sea lion management measures themselves have not been static. Statistically, changes that might be attributed to future management measures will be deeply "confounded" by changes in other components of the ecosystem.

To obtain targeted information requires definition of fisheries management options and specification of monitoring data that would allow determination of the efficacy of each option. By "efficacy" we mean a reversal of declining Steller sea lion abundance trends. This approach is referred to as an adaptive management experiment (Walters, 1986).

It remains unclear whether existing fisheries restrictions and closed areas should be treated as "precautionary" policies with respect to Steller sea lions. Some of the most credible hypotheses about the recent decline—including killer whale predation, continued illegal shooting and subsistence harvest, productivity declines due to regime shifts, and persistent changes in fish community structure due to predator-prey interactions—are not addressed by reducing the biomass of fish caught in sea lion critical habitat.

There are a few obvious data and monitoring requirements to include in future adaptive management options for Steller sea lions:

- Steller sea lion abundance, sex ratio, and survival/productivity rate from recapture patterns of marked (branded) animals at geographically diverse rookeries. Also, there should be continued monitoring of the response variables described in Chapter 6 at a variety of locations in the western and eastern stocks.
- Spatial and temporal patterns of fishing activity (efforts, fishing areas) for all fisheries working in proximity to Steller sea lions.
- Spatial abundance and seasonal distribution patterns of key fishes that are important prey for Steller sea lions in particular areas, including pollock, Atka mackerel, herring, and sand lance. In this regard, broader systematic use should be made of the localized depletion and tagging estimation procedures that have been tested recently in relation to Atka mackerel and the combined acoustic and trawl survey methods that have provided much insight into the details of interaction patterns in places like Prince William Sound.

**Recommendation: The NMFS should make a commitment to long-term sampling and monitoring of population and ecosystem variables such as sea lion and fish population structure and variation, harmful algal blooms, and oceanographic conditions indicative of climate regime shifts. Such intense monitoring should be continued until sea lions are delisted under the Endangered Species Act (ESA).**

**The fine geographic scale surveys of Steller sea lions on rookeries or haulout sites should be continued. Spatial analyses help distinguish causes that are global (or regional) in scale from causes that operate at local spatial scales. The documentation of site-specific trends, especially when evaluated in the context of related environmental variation, could greatly enhance understanding of the causes.**

**In addition, it is essential that the pup branding program be continued and expanded geographically. These data are essential for estimating pup and adult survival and would allow for the modernization of Steller sea lion demographics required for population models.**

**The observer program should be extended to smaller boats whenever feasible. Small boat fisheries (e.g., salmon, crab, herring) in the immediate vicinity of rookeries and haulouts should be monitored for interactions with sea lions, such as entanglement in or injury by fishing gear, disturbance of mothers and pups on rookeries, and illegal shooting or harassment. These activities may directly threaten sea lion survival relative to more gradual impacts associated with reduced availability of prey.**

Fishermen in Alaskan native and other communities constitute a largely untapped resource. Their involvement in sea lion research is essential from both an educational and a scientific perspective. Such endeavors could be modeled after the biosampling protocol used in the harbor seal program or the ongoing count of killer whales off Kodiak Island.

**Recommendation: More complete information should be obtained from the subsistence harvest of Steller sea lions. For example, tissues should be sampled, teeth extracted, sex determined, etc. Greater attention must be devoted to a socioeconomic study of all fishers and subsistence hunters, including efforts to enlist their knowledge of historic and contemporary Steller sea lion population trends.**

The possibility that killer whale predation is a major factor in the decline of the Steller sea lion population is not easily dismissed. Killer whales are abundant, intelligent, behaviorally flexible, and known to be rapacious consumers of large marine mammals. The predation of killer whales on Steller sea lions was previously thought to be minor, but recent reevaluations of their abundance suggest that killer whales could be a major source of sea lion mortality.

**Recommendation: An integrated set of investigations should be conducted to address the contribution of predation to sea lion mortality. Killer whale abundance is a critical component of all modeling approaches. Surveys should extend throughout the western and eastern ranges of Steller sea lions. In addition to abundance surveys, observational studies of killer whale feeding behavior will be important for estimating the size of the transient killer whale population. Transient animals are known to feed primarily on marine mammals such as sea lions, in contrast to resident killer whale pods that are believed to feed primarily on fish. Efforts should be made to determine if killer whales follow fishing vessels or are disproportionately aggregated around Steller sea lion breeding or haulout sites. Individual killer whales should be tagged so that their seasonal movements can be tracked by satellite. The resultant data may indicate whether individual whales (or their pods) are disproportionately associated with fishing operations or Steller sea lion rookeries and haulouts. A major advantage of satellite tags is that they reveal patterns of movement at locations inaccessible to observers, a particular problem during the harsh winter months.**

Slaughter of Steller sea lions for a wide variety of purposes—predator control, fox food, scientific inquiry—was sanctioned for decades before it was finally outlawed in 1990. What remains unknown is the extent to which the current law is obeyed. Loss of fish catches to sea lions was the leading motivation for historical predator control and eradication efforts. Although technology has improved the fish-finding and navigational abilities of fishing fleets, fishing gear (e.g., salmon gillnets, halibut longlines, groundfish trawls) remains largely unchanged, with continuing potential for interference by sea lions. Hence, the estimated unreported and illegal takes may be substantially in excess of the conservation modeling estimates.

**Recommendation: Confidential interviews and novel approaches should be used to document current levels of lost catch to Steller sea lions and contemporary responses of fishers, including moving to other fishing areas, use of seal bombs, and shooting of animals. The exercise will not be easy because protection of identity and confidentiality will be of paramount importance. Nonetheless, the**

**committee encourages efforts to document the extent of unauthorized killing of sea lions to help understand the unexplained Steller sea lion mortality.**

## RETROSPECTIVE ANALYSES

Because ecological systems are always in a state of flux, any "history" assumes a profound and confounding role in the reconstruction and understanding of specific population trajectories. In fact, the absence of adequate time series data has compromised both the population and ecosystem modeling efforts (see Chapter 6). Although a better understanding of the sea lion decline may not reveal why the western stock has failed to recover, two factors are particularly pertinent.

First, the role of infectious disease should be assessed through retrospective analyses of archival serum and tissue samples from 1980 to 1990 to cover the period of rapid population decline. At the same time, these samples should be tested for the presence of toxins and pollutants.

Second, fishery observer records should be consulted for accounts of killer whale attacks on sea lions. These records may constitute a valuable resource on plausible top-down influences on the sea lion population. Fishery observers should be tasked to document Steller sea lion sightings (killer whales and entanglements in fishing gear).

**Recommendation: A protocol should be implemented to test archival blood and tissue samples and collate observer records by year and geographic location.**

## MONITORING TO EVALUATE MANAGEMENT EFFICACY

The committee identified five possible fishery management options that should be paired with appropriate monitoring programs to provide information valuable for assessing prospects for reversing the population decline. These options are evaluated with regard to their scientific potential for discerning the role of the groundfish fishery in the Steller sea lion decline. Each option would require continuation of the existing monitoring program (i.e., continued census of trend sites and collection of demographic data based on pup branding and resighting). The committee made the assumption that it is possible to craft each of these options so as to satisfy the requirements of the ESA, but did not attempt to evaluate the legality of each option. The five options are presented below.

1. **Wait and see (maintain current closures indefinitely.** Current

closures offer one strategy to evaluate current trends in the Steller sea lion decline and the potential for recovery. Under this option, the most valuable monitoring information would be from tags that reveal immediate causes of mortality. A lack of recovery with the current closures would shift suspicion to one of the direct mortality hypotheses (continued culling, predators). A tagging program would provide some insight into the likelihood of these hypotheses so that management actions could be adjusted appropriately.

2. **Eliminate the *direct* fishery impacts, with greatly expanded closures.** This would mean closing most of the Atka mackerel fishery in the Aleutians, the main pollock fishing areas in the southern half of the eastern Bering Sea, the roe herring fisheries near rookeries in the Gulf of Alaska, and perhaps others. Under this option, the most important information needs would concern population dynamics of the fishes, particularly responses that may underlie the recent dominance of the fish community by gadoids. This community shift could become further entrenched if the groundfish catches are reduced through increasing restrictions on the fishery.
3. **Use replicated closed and open rookeries to experimentally evaluate localized fishery impacts.** Spatial management units consisting of two sets of closed and open areas would be established with each treatment area centered on a rookery. The western population would be divided into management regions with at least two closed and two open rookeries per region. Because most monitoring activities are conducted at rookeries (pup counts, measurement of vital rates, juvenile tagging, etc.) it makes the most sense to use rookeries (rather than rookeries and haulouts) as the experimental units. Also, sea lions are thought to be more vulnerable near rookeries because of the presence of pups and juveniles and because females must forage near the rookeries so they can return to nurse their pups. The closed treatment units would be subject to fishery closures and the open units would have sea lion-related fishery restrictions removed. The lifting of fishery restrictions in the open areas is important to achieve adequate contrast among treatments in the experiment.

   The logic in this design is to recognize that there is no best "baseline" or precautionary policy for the population in view of uncertainty about the causes of the decline so every part of the population should be treated as having a highly uncertain future no matter what restrictions are placed on the fisheries. Multiple (replicate) treatment/control sites are necessary for comparison because of the strong possibility that different portions of the sea lion population have and will continue to be affected by different stressors. If there are multiple problems due to multiple food/predation/fishing issues, replicated comparisons will be necessary to guard against incorrectly applying the results from any one treatment/control comparison to other areas that may not be comparable. Under this option, the most critical monitoring needs would be detailed local Steller sea lion censuses and spatial analysis of fish population changes for each experimental unit in the overall design.

4. **Implement a "titration experiment" that progressively increases restrictions on fisheries (such as area closures) until it is clear whether or not a positive response can be achieved via fishery management.** In essence, this option is a continuation of the historical policy on Steller sea lion protection since the ESA listing in 1992. While this "incrementalist" approach may be intuitively appealing from socioeconomic and practical management perspectives, it has at least two drawbacks. First, it would be taking place in a background of continuing ecosystem change due to factors such as oceanographic regime changes, so the "treatment effects" of management would be badly confounded with such changes on decadal timescales (for at least the next few decades). This would increase the chances of both "false positive" and "false negative" measured outcomes over the next 10 to 20 years at least. Second, this approach would have very high economic costs in the event of false positive signals that trigger the progression toward more fishery restrictions to continue for longer than they should. This economic cost would be in the form of lost fishery revenues and increased fishing costs for more distant operational deployment of fleets.
5. **Micromonitor and manage localized interactions between sea lions and fisheries to reduce mortality where and when it occurs in the future.** In this case, all basic monitoring activities (abundance, prey fields, mortality agent distribution) around at least some key rookeries would be increased to pinpoint the time and place of mortality events and take immediate management measures as necessary. This approach would require very considerable investment in invention and testing of new monitoring methods, particularly tagging systems for determining proximate causes of mortality. The expense would be greatly increased by a requirement for year-round, continuous monitoring to allow detection of concentrated impact in particular places or seasons. This approach is less able to detect and respond to evidence of delayed or cumulative interaction effects, such as increased predation in response to chronic food shortages. There is a disadvantage to commercial

fisheries because regulations might change frequently during the course of a fishing season.

**Recommendation: Of these management options, the committee concluded that option 3 offers the greatest benefits with regard to increasing the scientific understanding of the effect of fisheries on Steller sea lions. This is the only option that does not risk ineffective, long-term regulation of the fishing industry because it should reveal relatively quickly whether restrictions on fisheries will help prevent further decline. This option does not involve any pretense that the impacts of future changes in various "natural" factors like predator populations can eventually be unraveled through modeling and correlative studies. This option would provide a powerful comparative setting within which to carry out various detailed research studies on Steller sea lion behavior and performance in alternative ecological (prey field) environments while controlling for common effects of large-scale oceanographic regimes.**

The experimental policy option 3 is designed to improve management while at the same time facilitating research on the causes of decline and failure to recover. Open areas restore opportunities for fisheries by removing restrictions; closed areas remove potentially negative local impacts of fisheries on sea lions. A careful evaluation of past fishing effort in the proposed experimental areas will be required to assess the amount of displaced fishing effort. Placement of open areas where fishing effort has historically been high would decrease the potential for negative impacts arising from shifting effort from the closed to open areas. Some of the many considerations for the design of such an experiment include

- **Fished area (under normal management plans)**—The groundfish fisheries have been the focus of restrictions to protect sea lions based in part on the large amount of biomass removed by this fishery, but the potential effects of other fisheries have not been as thoroughly examined. Hence, there are two basic "experimental treatment" options for area closures: (1) closure to groundfish fisheries only, or (2) closure to all fishing. A positive response to treatment (1) would measure the impact of the groundfish fisheries separately from the effects of other fisheries. A positive response to treatment (2) would implicate fishing activities, but there would be uncertainty as to whether the response was due to exclusion of the groundfish fisheries or exclusion of another fishery, for example herring or salmon. Closure of these areas to all fishing activity would provide the greatest contrast with the open areas for assessment of fishery-related effects on Steller sea lions. If only the groundfish fisheries are excluded from the closed areas, logbook data and as much observer coverage as possible should be obtained for other fisheries. Closures should be designed to minimize displacement of fisheries to more distant, and less safe areas. Strict enforcement would be essential for the correct interpretation of effects of the closures.

- **Size and number of treatment areas**—The size of the closed areas depends on both fish movements and sea lion movements. The radius of the closure might range between 20 and 50 nautical miles (centered on a rookery). Replicates of each open/closed area comparison site will be required to assess the effects of environmental variability.
- **Timescale**—Some short-term results are possible (disease, possibly fish depletion, increase in direct mortality), but there must be recognition of and commitment to the need to maintain monitoring until longer-term recruitment and mortality responses are fully evident in the sea lion population (5 to 10 years). There should be a contingency plan to modify or curtail fishing in open areas if these areas show dramatic sea lion mortality as might occur under some scenarios, for example substantial fishery bycatch of Steller sea lions.

Although option #3 provides the best chance of providing new information about the role of the fisheries in the Steller sea lion decline, there is no guarantee that the outcome of this adaptive management experiment will provide an unequivocal result. A small, positive response to the fishery closures may be masked by fluctuations in other factors that also contribute to the decline, especially when the results are examined at shorter time scales. Nonetheless, this approach will at least indicate whether or not fishing activities near rookeries are a major factor threatening the recovery of the sea lion population and therefore will provide an improved baseline of knowledge for policymakers in the future.

# References


Aagaard, K., and E.C. Carmack. 1989. The role of sea ice and other fresh water in the Arctic circulation. *Journal of Geophysical Research* 94(C10):14485-14498.

Adams, T.C. 2000. Foraging differences and early maternal investment in adult female Alaskan Steller sea lions (*Eumetopias jubatus*). Ph.D. dissertation, Texas A&M University, Galveston.

Alaska Department of Fish & Game and National Marine Fisheries Service, Steller Sea Lion Research Program. 2001. Satellite telemetry and Steller sea lion research. Presentation to the National Research Council, Committee on the Alaska Groundfish Fishery and Steller Sea Lions, December.

Alaska Sea Grant College Program. 1993. *Is it Food?—Addressing Marine Mammal and Seabird Declines: Workshop Summary*. AK-SG-93-01. University of Alaska Fairbanks, Alaska Sea Grant College Program, Fairbanks.

Alverson, D.L. 1992. A review of commercial fisheries and the Steller sea lion *Eumetopias jubatus*: The conflict arena. *Reviews in Aquatic Sciences* 63(3-4):203-256.

Alverson, D.L., A.T. Pruter, and L.L. Ronholt. 1964. A study of demersal fishes and fisheries of the northeastern Pacific Ocean. H.R. MacMillan Lectures in Fisheries, Institute of Fisheries, University of British Columbia, Vancouver.

Anderson, P.J., and J.F. Piatt. 1999. Community reorganization in the Gulf of Alaska following ocean climate regime shift. *Marine Ecology Progress Series* 189:117-123.

Anderson, R.M. 1979. Parasite pathogenicity and the depression of host population equilibria. *Nature* 279(5709):150-152.

Andrews, R.D., D.G. Calkins, R.W. Davis, B.L. Norcross, K. Peijnenberg, and A.W. Trites. 2002. Foraging behavior and energetics of adult female Steller sea lions. Pp. 19-22 in *Steller Sea Lion Decline: Is It Food II?* D. DeMaster and S. Atkinson, eds. AK-SG-02-02. University of Alaska Fairbanks, Alaska Sea Grant College Program, Fairbanks.

Angliss, R.P., D.P. DeMaster, and A.L. Lopez. 2001. Alaska Marine Mammal Stock Assessments. 2001. U.S. Department of Commerce, NOAA Technical Memorandum, NMFS-AFSC-124. Washington, D.C.

Baduini, C.L., K.D. Hyrenbach, K.O. Coyle, A. Pinchuk, V. Mendenhall, and G.L. Hunt. 2001. Mass mortality of short-tailed shearwaters in the south-eastern Bering Sea during summer 1997. *Fisheries Oceanography* 10(1):117-130.

Bailey, K.M. 2000. Shifting control of recruitment of walleye pollock *Theragra chalcogramma* after a major climate and ecosystem shift. *Marine Ecology Progress Series* 198:215-224.

Barlough, J.E., E.S. Berry, E.A. Goodwin, R.F. Brown, R.L. DeLong, and A.W. Smith. 1987. Antibodies to marine calciviruses in the Steller sea lion (*Eumatopias jubatus schreber*). *Journal of Wildlife Diseases* 23(1):33-44.

Barrett-Lennard, L.G., K. Heise, E. Saulitis, G. Ellis, and C. Matkin. 1995. The impact of killer whale predation on Steller sea lion populations in British Columbia and Alaska. North Pacific Universities Marine Mammal Research Consortium, University of British Columbia, Vancouver. Available at *http://www.marinemammal.org/pdfs/Barrett_etal 1995-killer.pdf*.

Bartholomew, G.A. 1970. A model for the evolution of pinniped polygyny. *Evolution* 24(3):546-559.

Beamish, R.J., and D.R. Bouillon. 1993. Pacific salmon production trends in relation to climate. *Canadian Journal of Fisheries and Aquatic Sciences* 50(5):1002-1016.

Beamish, R.J., D.J. Noakes, G.A. McFarlane, L. Klyashtorin, V.V. Ivanov, and V. Kurashov. 1999. The regime concept and natural trends in the production of Pacific salmon. *Canadian Journal of Fisheries and Aquatic Sciences* 56(3):516-526.

Bean, T.H. 1887. The cod fishery of Alaska. Pp. 198-226 in *Fisheries and Fishery Industries of the U.S.*, Section 5, Vol. 2, History and Methods of the Fisheries, G.B. Goode, ed. U.S. Government Printing Office, Washington, D.C.

Beckmen, K.B., K.A. Burek, K.W. Pitcher, G.M. Ylitalo, and L.K. Lawrence. 2001. Organochlorine contaminants in live Steller sea lions, *Eumetopias jubatus*. P. 20 in *14th Biennial Conference on the Biology of Marine Mammals*, November 28-December 3, 2001, Vancouver, Canada.

Benson, A.J., and A.W. Trites. 2002. Ecological effects of regime shifts in the Bering Sea and eastern North Pacific Ocean. *Fish and Fisheries* 3:95-113.

Bickham, J.W., T.R. Loughlin, D.G. Calkins, J.K. Wickliffe, and J.C. Patton. 1998. Genetic variability and population decline in Steller sea lions from the Gulf of Alaska. *Journal of Mammalogy* 79(4):1390-1395.

Bigelow, H.B., and W.C. Shroeder. 1948. *Fishes of the Western North Atlantic. Part 1, Lancelets, Cyclostomes and Sharks*. Memorial Sears Foundation for Marine Research, Yale University, New Haven, Connecticut.

Blackburn, J., E. Brown, C. Brylinsky, D. Carlile, D. Chen, D. Clausen, D. Courtney, J. DiCosimo, M. Dorn, L. Fritz, J. Fujioka, S. Gaichas, M.A. Guttormsen, J. Heifetz, A. Hollowed, J. Ianelli, S. Lowe, C. Lunsford, B. Megrehy, V. O'Connell, M. Sigler, S. Stienessen, G. Thompson, B. Turnock, T. Wilderbuer, G. Williams, C. Wilson, and H. Zenger. 2001. *Stock Assessment and Fishery Evaluation Report for the Groundfish Resources of the Gulf of Alaska*. North Pacific Fisheries Management Council, Anchorage.

Bossart, G.D., D.G. Baden, R.Y. Ewing, B. Roberts, and S.D. Wright. 1998. Brevetoxicosis in manatees (*Trichechus manatus latirostris*) from the 1996 epizootic: Gross, histologic, and immunohistochemical features. *Toxicologic Pathology* 26(2):276-282.

Bowen, W.D., J. Harwood, D. Goodman, and G.L. Swartzman. 2001. Review of the November 2000 Biological Opinion and Incidental Take Statement with Respect to the Western Stock of the Steller Sea Lion, with Comments on the Draft August 2001 Biological Opinion. Final Report. North Pacific Fishery Management Council, Anchorage.

Braham, H.W., R.D. Everitt, and D.J. Rugh. 1980. Northern sea lion decline in the eastern Aleutian Islands. *Journal of Wildlife Management* 44(1):25-33.

Brandon, E.A.A. 2000. Maternal investment in Steller sea lions in Alaska. Ph.D. dissertation, Texas A&M University, Galveston.

Brandon, E.A.A., and R.W. Davis. 1999. Appendix B.5: Neonatal growth and condition and female attendance patterns. Pp. 20-22 in *Final Report on the Steller Sea Lion Physiology Workshop*, Seattle, February 8-10, 1999, A. Didier, ed. Pacific States Marine Fisheries Commission, Gladstone, Oregon.

Branson, J. 1971. Killer whales pursue sea lions in Bering Sea drama. *Commercial Fisheries Review* 33:39-40.

Bright, D.B. 1959. The occurrence and food of the sleeper shark, *Somniosis pacificus*, in a central Alaska bay. *Copea* 1959(1):76-77.

Brodeur, R.D., and D.M. Ware. 1992. Long-term variability in zooplankton biomass in the subarctic Pacific Ocean. *Fisheries Oceanography* 1(1):32-38.

Brodeur, R.D., C.E. Mills, J.E. Overland, G.E. Walters, and J.D. Schumacher. 1999. Evidence for a substantial increase in gelatinous zooplankton in the Bering Sea, with possible links to climate change. *Fisheries Oceanography* 8(4):296-306.

Brodeur, R.D., H. Sugisaki, and G.L. Hunt, Jr. 2002. Increases in jellyfish biomass in the Bering Sea: Implications for the ecosystem. *Marine Ecology Progress Series* 233:89-103.

Brooks, J.L., and S.I. Dodson. 1965. Predation, body size, and composition of plankton. *Science* 150:28-35.

Burek, K., R. Zarnke, T. Spraker, and A. Trites. 2001. Historical and current serology in Steller sea lions in Alaska. Pp. 35 in *Proceedings of the 14th Biennual Conference on the Biology of Marine Mammals*, Vancouver, Canada.

Calkins, D.G. 1985. Steller sea lion entanglement in marine debris. Pp. 308-314 in *Proceedings of the Workshop on the Fate and Impact of Marine Debris*, November 26-29, 1984, Honolulu, P.S. Shomura and H.O. Yoshida, eds. NOAA Technical Memorandum. NMFS-SWFC-54. U.S. Department of Commerce, National Oceanic and Atmospheric Administration, Washington, D.C.

Calkins, D.G., and E. Goodwin. 1988. *Investigation of the Declining Sea Lion Population in the Gulf of Alaska*. Alaska Department Fish & Game, Anchorage, Alaska.

Calkins, D.G., and K.W. Pitcher. 1982. *Population Assessment, Ecology and Trophic Relationships of Steller Sea Lions in the Gulf of Alaska*. Alaska Department Fish & Game, Anchorage.

Calkins, D.G., E.F. Becker, and K.W. Pitcher. 1998. Reduced body size of female Steller sea lions from a declining population in the Gulf of Alaska. *Marine Mammal Science* 14(2):232-244.

Callaway, D., K.R. Criddle, and E. Shea. 1998. Subsistence fisheries. Pp. 95-104 in *Implications of Global Change in Alaska and the Bering Sea Region*, G. Weller and P.A. Anderson, eds. Proceedings of a Workshop, Center for Global Change and Arctic System Research, University of Alaska, Fairbanks.

Carpenter, S.R., and J.F. Kitchell, eds. 1993. *The Trophic Cascade in Lakes*. Cambridge University Press, New York.

Castellini, M.A., and L.D. Rea. 1992. The biochemistry of natural fasting at its limits. *Experientia* 48(6):575-582.

Castellini, M.A., J.M. Castellini, and S.J. Trumble. 2000. *Recovery of Harbor Seals. Phase II: Controlled Studies of Health and Diet*. Restoration Project 01341. Alaska Department of Fish & Game, Habitat and Restoration Division, Exxon Valdez Oil Spill Trustee Council, Anchorage.

Castellini, M.A., D. Calkins, V. Burkanov, J.M. Castellini, S. Trumble, and T. Mau. 2001. Long-term feeding trials with harbor seals and Steller sea lions: Impact of different diets on health and nutrition. P. 40 in *14th Biennial Conference on the Biology of Marine Mammals*. Vancouver, Canada.

Clark, W.G. 1991. Groundfish exploitation rates based on life history parameters. *Canadian Journal of Fisheries and Aquatic Sciences* 48(5):734-750.

Cooper, L.W., J.M. Grebmeier, I.L. Larsen, V.G. Egorov, C. Theodorakis, H.P. Kelly, and J.R. Lovvorn. 2002. Seasonal variation in sedimentation of organic materials in the St. Lawrence Island polynya region, Bering Sea. *Marine Ecology Progress Series* 226:13-26.

Criddle, K.R., M. Herrmann, J.A. Greenberg, and E.M. Feller. 1998. Climate fluctuations and revenue maximization in the eastern Bering Sea fishery for walleye pollock. *North American Journal of Fisheries Management* 18:1-10.

Dailey, M.D. 2001. Parasitic diseases. Pp. 357-379 in *CRC Handbook of Marine Mammal Medicine*, 2nd ed., L.A. Dierauf and F.M.D Gulland, eds. CRC Press, Boca Raton, Florida.

D'Arrigo, R., R. Villalba, and G. Wiles. 2001. Tree-ring estimates of Pacific decadal climate variability. *Climate Dynamics* 18(3-4):219-224.

Davis, R.W., E.A.A. Brandon, T.C. Adams, T.M Williams, M.A. Castellini, T.R. Loughlin, and D.G. Calkins. 1996. Indices of reproductive effort, body condition and pup growth for Steller sea lions (*Eumetopias jubatus*) in Alaska. Pp. 53-61 in *Steller Sea Lion Recovery Investigations in Alaska, 1992-1994*, K. Pitcher, ed. Alaska Department of Fish & Game, Division of Wildlife Conservation, Wildlife Technical Bulletin, Anchorage.

DeMaster, D., and S. Atkinson, eds. 2002. *Steller Sea Lion Decline: Is It Food II*? University of Alaska Sea Grant Program, Fairbanks.

DeMaster, D., and L. Fritz. 2001. FY01 Steller Sea Lion—Groundfish Budget $43.15 Million. Meeting handout. National Research Council, Committee on the Alaska Groundfish Fishery and Steller Sea Lions, August, Seattle.

deSwart, R.L., P.S. Ross, L.J. Vedder, H.H. Timmerman, S.H. Heisterkamp, H. van Loveren, J.G. Vos, P.J.H. Reijnders, and A.D.M.E. Osterhaus. 1994. Impairment of immunological functions in harbor seals (*Phoca vitulina*) feeding on fish from polluted coastal waters. *Ambio* 23(2):155-159.

DiCosimo, J. 1998. A summary of assessment information for managing Alaska groundfish stocks. Pp. 841-851 in *Proceedings of the International Symposium on Fishery Stock Assessment Models for the 21st Century*. F. Funk et al., eds. University of Alaska, Alaska Sea Grant College Program, Fairbanks.

Dierauf, L.A., and F.M.D. Gulland, eds. 2001. *CRC Handbook of Marine Mammal Medicine*, 2nd ed. CRC Press, Boca Raton, Florida.

Dorn, M.W., A.B. Hollowed, E. Brown, B. Megrey, C. Wilson, and J. Blackburn. 2001. Assessment of the walleye pollock stock in the Gulf of Alaska. Pp. 1-1–1-90 in *Stock Assessment and Fishery Evaluation Report for the Groundfish Resources of the Gulf of Alaska*, J. Blackburn et al., eds. North Pacific Fishery Management Council, Anchorage.

Dragoo, E.E., G.V. Byrd, Jr., and D.B. Irons. 2000. Breeding status and population trends of seabirds in Alaska in 1999. U.S. Department of Interior, Fish and Wildlife Service, Washington, D.C.

Dunn, J.L., J.D. Buck, and T.R. Robeck. 2001. Bacterial diseases of cetaceans and pinnipeds. Pp. 309-335 in *CRC Handbook of Marine Mammal Medicine*, 2nd ed., L.A. Dierauf and F.M.D. Gulland, eds. CRC Press, Boca Raton, Florida.

Eberhardt, L.L. 1977. "Optimal" management policies for marine mammals. *Wildlife Society Bulletin* 5:162-169.

Elliott, H.W. 1887. The sea lion hunt. Pp. 467-474 in *The Fisheries and Fishery Industries of the U.S.*, Sec. 5, Vol. 2, G.B. Goode, ed. U.S. Government Printing Office, Washington, D.C.

Elton, C.S. 1927. *Animal Ecology*. Sidgwick & Jackson, Ltd., London.

Emlen, J.M., and M.G.R. Emlen. 1975. Optimal choice in diet: Test of a hypothesis. *American Naturalist* 109(968):427-435.

Estes, J.A., and J.F. Palmisano. 1974. Sea otters: Their role in structuring nearshore communities. *Science* 185:1058-1060.

Estes, J.A., M.T. Tinker, T.M. Williams, and D.F. Doak. 1998. Killer whale predation on sea otters linking coastal and nearshore ecosystems. *Science* 282(5388):473-476.
Estes, J.A., K. Crooks, and R. Holt. 2001. Predators, ecological role of. Pp. 857-878 in *Encyclopedia of Biodiversity*, S.A. Levin, ed. Academic Press, San Diego.
Ferrero, R.C., and L.W. Fritz. 2002. Steller Sea Lion Research and Coordination: A Brief History and Summary of Recent Progress. NOAA Technical Memorandum NMFS-AFSC-129. U.S. Department of Commerce, National Oceanic and Atmospheric Administration, National Marine Fisheries Service, Alaska Fisheries Science Center, Anchorage.
Ferrero, R.C., D.P. DeMaster, P.S. Hill, M.M. Muto, and A.L. Lopez. 2000. Alaska Marine Mammal Stock Assessments, 2000. NOAA Technical Memorandum NMFS-AFSC-119. U.S. Department of Commerce, National Oceanic and Atmospheric Administration, Anchorage.
Finney, B.P., I. Gregory-Eaves, J. Sweetman, M.S.V. Douglas, and J.P. Smol. 2000. Impacts of climate change and fishing on Pacific salmon abundance over the past 300 years. *Science* 290:795-799.
Fiscus, C.H., and G.A. Baines. 1966. Food and feeding behavior of Steller and California sea lions. *Journal of Mammalogy* 42(2):195-200.
Fletcher, D., F.M.D. Gulland, M. Haulena, L.J. Lowenstine, and M. Dailey. 1998. Nematode-associated gastrointestinal perforations in stranded California sea lions (*Zalophus californianus*). Pp. 59-60 in *Proceedings of the 29th International Association for Aquatic Animal Medicine Conference*, San Diego. International Association for Aquatic Animal Medicine, Moscow.
Ford, J.K.B., and G.M. Ellis. 1999. *Transients: Mammal-Hunting Killer Whales of British Columbia, Washington, and Southeastern Alaska*. University of British Columbia Press, Vancouver.
Ford, J.K.B., G.M. Ellis, L.G. Barrett-Lennard, A.B. Morton, R.S. Palm, and K.C. Balcomb III. 1998. Dietary specialization in two sympatric populations of killer whales (*Orcinus orca*) in coastal British Columbia and adjacent waters. *Canadian Journal of Zoology* 76:1456-1471.
Francis, R.C., and S.R. Hare. 1994. Decadal-scale regime shifts in the large marine ecosystem of the North-east Pacific: A case of historical science. *Fisheries Oceanography* 3(4):279-291.
Francis, R.C., S.R. Hare, A.B. Hollowed, and W.S. Wooster. 1998. Effects of interdecadal climate variability on the oceanic ecosystems of the NE Pacific. *Fisheries Oceanography* 7(1):1-21.
Fritz, L.W. 1993. Trawl Locations of Walleye Pollock and Atka Mackerel Fisheries in the Bering Sea, Aleutian Islands and Gulf of Alaska From 1977-92. AFSC Processed Report 93-08. U.S. Department of Commerce, National Oceanic and Atmospheric Administration, National Marine Fisheries Service, Seattle.
Fritz, L.W. 1999. Do trawl fisheries off Alaska create localized depletions of atka mackerel (*Pleurogrammus monopterygius*)? Appendix 1 in *Environmental Assessment/Regulatory Impact Review and Final Regulatory Flexibility Analysis for a Regulatory Amendment to Reapportion Total Allowable Catch of Atka Mackerel and Reduce Effects on Steller Sea Lions*. National Marine Fisheries Service, Alaska Fisheries Science Center, Juneau.
Fritz, L. 2001. North Pacific groundfish, groundfish fisheries, and Steller sea lions. Presentation to the National Research Council, Committee on the Alaska Groundfish Fishery and Steller Sea Lions, Seattle, December.

Fritz, L.W., A. Greig, and R. Reuter. 1998. Catch-Per-Unit-Effort, Length, and Depth Distributions of Major Groundfish and Bycatch Species in the Bering Sea, Aleutian Islands and Gulf of Alaska Regions Based on Groundfish Fishery Observer Data. NOAA Technical Memorandum NMFS-AFSC-88. U.S. Department of Commerce, National Oceanic and Atmospheric Administration, National Marine Fisheries Service, Seattle.

Fritz, L.W., S.F. McDermott, and S.A. Lowe. 2001. Efficacy of trawl fishery exclusion zones in maintaining prey availability for Steller sea lions: Description of Atka mackerel tagging project in Seguam Pass, Aleutian Islands, Alaska, in 1999 and 2000. Pp. 75-87 in *Steller Sea Lion Investigations*, B.S. Fadely, ed. AFSC Processed Report 2001-05, National Marine Fisheries Service, Alaska Fishery Science Center, Seattle.

Funk, F., and K.A. Rowell. 1995. Population model suggests new threshold for managing Alaska's Togiak fishery for Pacific herring in Bristol Bay. *Alaska Fishery Research Bulletin* 2(2):125-126.

Gedalof, Z., and D.J. Smith. 2001. Interdecadal climate variability and regime-scale shifts in Pacific North America. *Geophysical Research Letters* 28(8):1515-1518.

Gentry, R.L. 2002. Eared seals. Pp. 348-351 in *Encyclopedia of Marine Mammals*, W.F. Perrin, B.G. Würsig, and J.G.M. Thewissen, eds. Academic Press, San Diego.

Gerber, L.R., and G.R. VanBlaricom. 2001. Implications of three viability models for the conservation status of the western population of Steller sea lions (*Eumetopias jubatus*). *Biological Conservation* 102(3):261-269.

Grebmeier, J.M. 1993. Studies on pelagic benthic coupling extended onto the Russian continental shelf in the northern Bering and Chukchi Seas. *Continental Shelf Research* 13(5-6):653-668.

Grebmeier, J.M., and L.W. Cooper. 2002. Benthic processes in the Bering Strait region of the Arctic: Temporal/spatial variability and global change. Ocean Sciences Meeting Supplement, Abstract OS32M-01. *EOS, Transactions of the American Geophysical Union* 83(4):268.

Grebmeier, J.M., and K.H. Dunton. 2000. Benthic processes in the northern Bering/Chukchi Seas: Status and global change. Pp. 80-93 in *Impacts of Changes in Sea Ice and Other Environmental Parameters in the Arctic,* H.P. Huntington, ed. Proceedings of a Marine Mammal Commission Workshop, February 15-17, Girdwood, Alaska.

Grenfell, B.T., M.E. Lonergan, and J. Harwood. 1992. Quantitative investigations of the epidemiology of phocine distemper virus (PDV) in European common seal populations. *The Science of the Total Environment* 115:15-29.

Groom, M., and M. Pascual. 1998. The analysis of population viability: An outlook on the practice of viability analysis. Pp. 4-27 in *Conservation Biology for the Coming Decade*, P. Fielder and P. Kareiva, eds. Chapman and Hall, New York.

Gulland, F.M.D. 1992. The role of nematode parasites in Soay sheep (*Ovis-aries* L.) mortality during a population crash. *Parasitology* 105:493-503.

Hairston, N.G., F.E. Smith, and L.B. Slobodkin. 1960. Community structure, population control, and competition. *American Naturalist* 94:421-425.

Hanni, K.D., J.A.K. Mazet, F.M.D. Gulland, J. Estes, M. Staedler, M.K. Murray, and D. Jessup. In press. Comparison of clinical pathological values and pathogen exposure between California and Alaskan sea otters. *Journal of Wildlife Diseases.*

Hanski, I., and D. Simberloff. 1997. The metapopulation approach, its history, conceptual domain, and application to conservation. Pp. 5-26 in *Metapopulation Biology*, I.A. Hanski and M.E. Gilpin, eds. Academic Press, San Diego.

Hare, S.R., and R.C. Francis. 1994. Climate change and salmon production in the northeast Pacific Ocean. Pp. 357-372 in *Climate Change and Northern Fish Populations*, R.J. Beamish, ed. Canadian Special Publication in Fisheries and Aquatic Sciences 121.

Hare, S.R., and N.J. Mantua. 2000. Empirical evidence for North Pacific regime shifts in 1977 and 1989. *Progress in Oceanography* 47(2-4):103-145.

Hare, S.R., N.J. Mantua, and R.C. Francis. 1999. Inverse production regimes: Alaskan and West Coast Pacific Salmon. *Fisheries* 21(1):6-14.

Harmon, H.L. 2001. Seasonal reproductive endocrinology and anatomy of Steller sea lions (*Eumetopias jubatus*). M.S. thesis. University of Alaska, Fairbanks.

Harwood, J., and A. Hall. 1990. Mass mortality in marine mammals: Its implications for population dynamics and genetics. *Trends in Ecology & Evolution* 5(8):254-257.

Haynes, T.L., and C. Mishler. 1994. The subsistence harvest and use of Steller sea lions in Alaska. Technical Paper 198. Alaska Department of Fish & Game, Division of Subsistence.

Heesterbeek, J.A.P., and M.G. Roberts. 1995. Mathematical models of microparasites of wildlife. Pp. 90-122 in *Ecology of Infectious Diseases in Natural Populations*, B.T. Grenfell and A.P. Dobson, eds. Cambridge University Press, Cambridge, U.K.

Heppell, S.S., H. Caswell, and L.B. Crowder. 2000. Life histories and elasticity patterns: Perturbation analysis for species with minimal demographic data. *Ecology* 81(3):654-665.

Hernández, M., I. Robinson, A. Aguilar, L.M. González, L.F. López-Jurado, M.I. Reyero, E. Cacho, J. Franco, V. López-Rodas, and E. Costas. 1998. Did algal toxins cause monk seal mortality? *Nature* 393(6680):28-29.

Hirons, A.C., D.M. Schell, and B.P. Finney. 2001. Temporal trends of $\delta^{13}C$ and $\delta^{15}N$ in North Pacific pinnipeds: Inferences regarding environmental change and diet. *Oecologia* 129(4):591-601.

Holling, C.S. 1973. Resilience and stability of ecological systems. *Annual Review of Ecology and Systematics* 4:1-23.

Hollowed, A.B., and W.S. Wooster. 1992. Variability of winter ocean conditions and strong year classes of Northeast Pacific groundfish. *ICES Marine Science Symposium* 195:433-444.

Hollowed, A.B., and W.S. Wooster. 1995. Decadal-scale variations in the eastern Subarctic Pacific: II. Response of Northeast Pacific fish stocks. Pp. 373-385 in *Climate Change and Northern Fish Populations*, R.J. Beamish, ed. *Canadian Journal of Fisheries and Aquatic Sciences* 121(Special Publication).

Hollowed, A.B., S.R. Hare, and W.S. Wooster. 1998. Pacific basin climate variability and patterns of Northeast Pacific marine fish production. Pp. 89-104 in *Proceedings of the 10th 'Aha Huliko'a Hawaiian Winter Workshop on Biotic Impacts of Extratropical Climate Variability in the Pacific*, University of Hawaii at Manoa, January 26-29, G. Holloway, P. Muller, and D. Henderson, eds. NOAA Award No. NA67RJ0154, SOEST Special Publication.

Hoover, A.A. 1988. Steller sea lion. Pp. 159-193 in *Selected Marine Mammals of Alaska: Species Accounts with Research and Management Recommendations*, J.W. Lentfer, ed. Marine Mammal Commission, Washington, D.C.

Hourston, A.S., and C.W. Haegele. 1980. Herring on Canada's Pacific coast. *Canadian Journal of Fisheries and Aquatic Sciences* 48(Special Publication).

Hu, F.S., D. Slawinski, H.E. Wright, Jr., E. Ito, R.G. Johnson, K.R. Kelts, R.F. McEwan, and A. Boedigheimer. 1999. Abrupt changes in North American climate during early Holocene times. *Nature* 400:437-440.

Hudson, P., and A.P. Dobson. 1995. Macroparasites: Observed patterns in naturally fluctuating animal populations. Pp. 144-176 in *Ecology of Infectious Diseases in Natural Populations*, B.T. Grenfell and A.P. Dobson, eds. Cambridge University Press, Cambridge, U.K.

Hunt, G.L., Jr., P. Stabeno, G. Walters, E. Sinclair, R.D. Brodeur, J.M. Napp, and N.A. Bond. 2002. Climate change and control of the southeastern Bering Sea pelagic ecosystem. *Deep-Sea Research Part II-Topical Studies in Oceanography* 49(26):5821-5853.

Hunter, M.D., and P.W. Price. 1992. Playing chutes and ladders: Heterogeneity and the relative roles of bottom-up and top-down forces in natural communities. *Ecology* 73(3):724-732.

Ianelli, J.N., T. Buckley, T. Honkalehto, G. Walters, and N. Williamson. 2001. Eastern Bering Sea pollock assessment. Pp. 1.1-1.89 in *Stock Assessment and Fishery Evaluation Report for the Groundfish Resources of the Bering Sea/Aleutian Islands Regions*. North Pacific Fishery Management Council, Anchorage.

Imler, R.H., and H.R. Sarber. 1947. *Harbor Seals and Sea Lions in Alaska*. Special Scientific Report 28. U.S. Department of the Interior, Fish and Wildlife Service, Washington, D.C.

Ingraham, W.J., Jr., C.C. Ebbesmeyer, and R.A. Hinrichsen. 1998. Imminent climate and circulation shift in northeast Pacific Ocean could have major impact on marine resources. *EOS, Transactions of the American Geophysical Union* 79(16):197, 199, 201.

Iverson, S.J. 2001. Estimating marine mammal diets using quantitative fatty acid signature analysis (QFASA). P. 106 in *14th Biennial Conference on the Biology of Marine Mammals, November 28-December 3, 2001*. Vancouver, Canada.

Jackson, J.B.C., M.X. Kirby, W.H. Berger, K.A. Bjorndal, L.W. Botsford, B.J. Bourque, R.H. Bradbury, R. Cook, J. Erlandson, J.A. Estes, T.P. Hughes, S. Kidwell, C.B. Lange, H.S. Lenihan, J.M. Pandolfi, C.H. Peterson, R.S. Steneck, M.J. Tegner, and R.R. Warner. 2001. Historical overfishing and the recent collapse of coastal ecosystems. *Science* 293(5530):629-638.

Jensen, T., M. van de Bildt, H.H. Dietz, T.H. Andersen, A.S. Hammer, T. Kuiken, and A. Osterhaus. 2002. Another phocine distemper outbreak in Europe. *Science* 297:209.

King, J.E. 1983. *Seals of the World*, 2nd ed. University of Queensland Press, St. Lucia.

Kowalik, Z. 1999. Bering Sea tides. Pp. 93-127 in *Dynamics of the Bering Sea: A Summary of Physical, Chemical, and Biological Characteristics, and a Synopsis of Research in the Bering Sea.*, T.R. Loughlin and K. Ohtani, eds. University of Alaska, Alaska Sea Grant College Program, Fairbanks.

Krebs, J.R. 1978. Optimal foraging: Decision rules for predators. Pp. 23-63 in *Behavioural Ecology*, J.R. Krebs and N.B. Davies, eds. Blackwell Scientific Publications, Oxford.

Kruse, G.H. 1998. Salmon run failures in 1997-1998: A link to anomalous ocean conditions? *Alaska Fishery Research Bulletin* 5(1):55-63.

Kruse, G.H., F.C. Funk, H.J. Geiger, K.R. Mabry, H.M. Savikko, and S.M. Siddeek. 2000. Overview of State-Managed Marine Fisheries in the Central and Western Gulf of Alaska, Aleutian Islands, and Southeastern Bering Sea, with Reference to Steller Sea Lions. Regional Information Report 5J00-10. Alaska Department of Fish & Game, Division of Commercial Fisheries, Juneau.

Kruse, G.H., M. Crow, E.E. Krygier, D.S. Lloyd, K.W. Pitcher, L.D. Rea. M. Ridgway, R.J. Small, J. Stinson, and K.M. Wynne. 2001. A Review of Proposed Fishery Management Actions and the Decline of Steller Sea Lions *Eumetopias Jubatus* in Alaska: A Report by the Steller Sea Lion Restoration Team. ADF&G Regional Information Report No. 5J01-04. Alaska Department of Fish & Game, Division of Commercial Fisheries, Juneau.

Laevastu T., F. Favorite, and W.B. McAlister. 1976. Dynamic marine ecosystem model for evaluation of marine resources in the Bering Sea. *Proceedings of the Alaskan Science Conference, American Association for the Advancement of Science* 27:180-181.

Laevastu, T., F. Favorite, and H.A. Larkins. 1982. Resource assessment and evaluation of the dynamics of the fisheries resources in the northeastern Pacific with numerical ecosystem models. In Multispecies Approaches to Fisheries Management Advice, M.C. Mercer, ed. *Special Publication of Canadian Journal of Fisheries Aquatic Sciences* 59:70-81.

Laughlin, W.S. 1980. *Aleuts: Survivors of the Bering Land Bridge*. Holt, Rinehart & Winston, New York.

Le Boeuf, B.J. 1994. Variation in the diving pattern of northern elephant seals with age, mass, sex and reproductive condition. Pp. 237-251 in *Elephant Seals: Population Ecology, Behavior and Physiology*, B.J. Le Boeuf, and R.M. Laws, eds. University of California Press, Berkeley.

Lee, J.S., S. Tanabe, H. Umino, R. Tatsukawa, T.R. Loughlin, and D.C. Calkins. 1996. Persistent organochlorines in Steller sea lion (*Eumatopias jubatus*) from the bulk of Alaska and the Bering Sea 1976-1981. *Marine Pollution Bulletin* 32(7):535-544.

Lewontin, R.C. 1969. The meaning of stability. Pp. 13-24 in *Diversity and Stability in Ecological Systems-Report of Symposium Held May 26-28, 1969*. Brookhaven Symposium in Biology No. 22. Brookhaven National Laboratory, Upton, New York.

Livingston, P.A., and R.D. Methot. 1998. Incorporation of predation into a population assessment model of eastern Bering Sea walleye pollock. Pp. 663-678 in *Fishery Stock Assessment Models*. NOAA Technical Report 126, NMFS/NWC-54. Alaska Sea Grant College Program, Fairbanks.

Livingston, P.A., L.L. Low, and R.J. Marasco. 1999. Eastern Bering Sea ecosystem trends. Pp. 140-162 in *Large Marine Ecosystems of the Pacific Rim: Assessment, Sustainability, and Management*, Q. Tang and K. Sherman, eds. Blackwell Science, Inc., Malden, Massachusetts.

Loughlin, T.R. 1987. Report of the Workshop on the Status of Northern Sea Lions in Alaska Held December 9 and 10, 1986. NWAFC/NMFS/NOAA. Processed Report No. 87-04. National Marine Fisheries Service, National Oceanic and Atmospheric Administration, National Marine Fisheries Service, Seattle.

Loughlin, T.R. 1998. The Steller sea lion: A declining species. *Biosphere Conservation* 1(2):91-98.

Loughlin, T.R., and R.L. DeLong. 1983. Incidental Catch of Northern Sea Lions During the 1982 and 1983 Walleye Pollock Joint Venture Fishery, Shelikof Strait, Alaska. NWAFC Processed Report 83-15. Northwest and Alaska Fishery Science Center, National Marine Fisheries Service, National Oceanic and Atmospheric Administration, U.S. Department of Commerce, Washington, D.C.

Loughlin, T.R., and R. Nelson, Jr. 1986. Incidental mortality of northern sea lions in Shelikof Strait, Alaska. *Marine Mammal Science* 2(1):14-33.

Loughlin, T.R., and A.E. York. 2000.[1] An accounting of the sources of Steller sea lion (*Eumetopias jubatus*) mortality. *Marine Fisheries Review* 62:40-45.

Loughlin, T.R., and A.E. York. 2001. Northern fur seal (*Callorhinus ursinus*) and Steller sea lion (*Eumetopias jubatus*) status and trends. Presentation given by T.R. Loughlin to the National Research Council, Committee on the Alaska Groundfish Fishery and Steller Sea Lions, Seattle, December.

Loughlin, T.R., L. Consiglieri, R.L. DeLong, and A.T. Actor. 1983. Incidental catch of marine mammals by foreign fishing vessels, 1978-81. *Marine Fisheries Review* 45(7-9):44-49.

Loughlin, T.R., D.J. Rugh, and C.H. Fiscus. 1984. Northern sea lion distribution and abundance: 1956-1980. *Journal of Wildlife Management* 48(3):729-740.

[1]Please note that this article in *Marine Fisheries Review* was printed in 2002 although the publication date on the journal is listed as 2000.

Loughlin, T.R., P.J. Gearin, R.L. DeLong, and R.L. Merrick. 1986. Assessment of Net Entanglement on Northern Sea Lions in the Aleutian Islands, 25 June-15 July 1985. National Marine Mammal Laboratory, Seattle.

Loughlin, T.R., A.S. Perlov, and V.V. Vladimirov. 1992. Range-wide survey and estimation of total number of Steller sea lions in 1989. *Marine Mammal Science* 83:220-239.

Lovvorn, J.R., C.L. Baduini, and G.L. Hunt. 2001. Modeling underwater visual and filter feeding by planktivorous shearwaters in unusual sea conditions. *Ecology* 82(8):2342-2356.

Lovvorn, J.R., S.E. Richman, J.M. Grebmeier, and L.W. Cooper. In press. Diet and body condition of spectacled eiders wintering in pack ice of the Bering Sea. *Polar Biology*.

Lowe, S.A., R.F. Reuter, and H. Zenger. 2001. Assessment of the Bering Sea/Aleutian Islands Atka mackerel. Pp.13-1 to 13-44 (plus 17 figures) in *Stock Assessment and Fishery Evaluation Report for the Groundfish Resources of the Bering Sea/Aleutian Islands Regions*. North Pacific Fishery Management Council. Anchorage.

Lynch, A.H., W.L. Chapman, J.E. Walsh, and G. Weller. 1995. Development of a regional climate model of the western Arctic. *Journal of Climate* 8(6):1555-1570.

MacCall, A.D. 1990. *Dynamic Geography of Marine Populations*. University of Washington Press, Seattle.

Mace, P.M. 1994. Relationships between common biological reference points used as thresholds and targets of fisheries management strategies. *Canadian Journal of Fisheries and Aquatic Sciences* 51(1):110-122.

Manabe, S., R.J. Stouffer, M.J. Spelman, and K. Bryan. 1991. Transient responses of a coupled ocean-atmosphere model to gradual changes of atmospheric $CO_2$. Part I: Annual mean response. *Journal of Climate* 4(8):785-818.

Mantua, N.J., and S.R. Hare. 2002. The Pacific Decadal Oscillation. *Journal of Oceanography* 58(1):35-44.

Mantua, N.J., S.R. Hare, Y. Zhang, J.M. Wallace, and R.C. Francis. 1997. A Pacific interdecadal oscillation with impacts on salmon production. *Bulletin of the American Meteorological Society* 78(6):1069-1079.

Maser, C., B.R. Mate, J.F. Franklin, and C.T. Dyrness. 1981. Natural History of Oregon Coast Mammals. General Report PNW-133. U.S. Department of Interior, Bureau of Land Management, Washington, D.C.

Mathisen, O.A. 1959. Studies on Steller sea lions (*Eumetopias jubatus*) in Alaska. In *Transactions of the North American Wildlife and Natural Resources Conference* 24:346-356.

Mathisen, O.A., R.T. Baade, and R.J. Lopp. 1962. Breeding habits, growth and stomach contents of the Steller sea lion in Alaska. *Journal of Mammalogy* 43(4):469-477.

Matkin, C.O. 1977. A Preliminary evaluation of fishery depredation by marine mammals on the Copper River Flats, Alaska, spring 1977. University of Alaska, Graduate Wildlife Office, College. Available at: *http://research.alaskasealife.org/*.

Matkin, C.O., and F.H. Fay. 1980. Marine Mammal-Fishery Interactions on the Copper River and in Prince William Sound, Alaska, 1978. Marine Mammal Commission, Washington, D.C.

Matkin, C.O., G.M. Ellis, E.I. Saulitis, L.G. Barrett-Lennard, and D. Matkin. 1999. Killer Whales of Southern Alaska. North Gulf Oceanic Society, Homer, Alaska.

May, R.M. 1977. Thresholds and breakpoints in ecosystems with a multiciplicity of stable states. *Nature* 269(5628):471-477.

Menge, B.A. 1995. Indirect effects in marine rocky intertidal interaction webs: Patterns and importance. *Ecological Monographs* 65(1):21-74.

Merrick, R.L., and D.G. Calkins. 1996. Importance of juvenile walleye pollock, *Theragra chalcogramma*, in the diet of Gulf of Alaska Steller sea lions, *Eumetopias jubatus*. Pp. 153-166 in *Ecology of Juvenile Walleye Pollock (Theragra chalcogramma)*, R.D. Brodeur et al., eds. NOAA/NMFS Technical Report 126. U.S. Department of Commerce, National Oceanic and Atmospheric Administration, Washington, D.C.

Merrick, R.L., and T.R. Loughlin. 1997. Foraging behavior of adult female and young-of-the-year Steller sea lions in Alaskan waters. *Canadian Journal of Zoology* 75(5):776-786.

Merrick, R.L., T.R. Loughlin, and D.G. Calkins. 1987. Decline in abundance of the northern sea lion, *Eumetopias jubatus*, in Alaska, 1956-86. *Fishery Bulletin* 85(2):351-365.

Merrick, R.L., R. Brown, D.G. Calkins, and T.R. Loughlin. 1995. A comparison of Steller sea lion, *Eumetopias jubatus*, pup masses between rookeries with increasing and decreasing populations. *Fishery Bulletin* 93(4):753-758.

Merrick, R.L., M.K. Chumbley, and G.V. Byrd. 1997. Diet diversity of Steller sea lions (*Eumetopias jubatus*) and their population decline in Alaska: A potential relationship. *Canadian Journal of Fisheries and Aquatic Sciences* 54(6):1342-1348.

Methot, R.D. 2000. Technical Description of the Stock Synthesis Assessment Program. NOAA Technical Memorandum NMFS-NWFSC-43. U.S. Department of Commerce, National Oceanic and Atmospheric Administration, National Marine Fisheries Service, Northwest Fisheries Science Center, Seattle.

Meyers, T.R., S. Short, K. Lipson, W.N. Batts, J.R. Winton, J. Wilcock, and E. Brown. 1994. Association of viral hemorrhagic septicemia virus with epizootic hemorrhages of the skin in Pacific herring *Clupea harengus pallasi* from Prince William Sound and Kodiak Island, Alaska, USA. *Diseases of Aquatic Organisms* 19(1):27-37.

Milette, L.L. 1999. Behaviour of lactating Steller sea lions (*Eumetopias jubatus*) during the breeding season: A comparison between a declining and stable population in Alaska. M.S. thesis, University of British Columbia.

Minobe, S. 1997. A 50-70 year oscillation over the North Pacific and North America. *Geophysical Research Letters* 24(6):683-686.

Minobe, S. 2002. Interannual to interdecadal changes in the Bering Sea and concurrent 1998/99 changes over the North Pacific. *Progress in Oceanography* 55(1-2):45-64.

Mueter, F.J., and B.L. Norcross. 2000. Changes in species composition of the demersal fish community in nearshore waters of Kodiak Island, Alaska. *Canadian Journal of Fisheries Aquatic Sciences* 57(6):1169-1180.

National Marine Fisheries Service. 1999. Endangered Species Act—Section 7 Consultation—Biological Opinion, draft dated December 22. U.S. Department of Commerce, National Oceanic and Athmospheric Administration, National Marine Fisheries Service, Juneau. Available at *http://www.fakr.noaa.gov/protectedresources/akground2000.pdf*

National Marine Fisheries Service. 2000. Endangered Species Act—Section 7 Consultation. Biological Opinion and Incidental Take Statement. November 30. U.S. Department of Commerce, National Oceanic and Atmospheric Administration, National Marine Fisheries Service, Juneau.

National Marine Fisheries Service. 2001. *Steller Sea Lion Protection Measures, Draft Supplemental Environmental Impact Statement, Volume II.* U.S. Department of Commerce, National Oceanic and Atmospheric Administration, National Marine Fisheries Service, Alaska Region, Juneau.

National Marine Fisheries Service. 2002. Toward Rebuilding America's Fisheries. Annual Report to Congress on the Status of U.S. Fisheries—2001. U.S. Department of Commerce, National Oceanic and Atmospheric Administration, National Marine Fisheries Service, Office of Sustainable Fisheries, Silver Spring. Available at *http://www.nmfs.noaa/gov/sfa/reports.html* [June 26, 2002].

National Research Council. 1996. *The Bering Sea Ecosystem*. National Academy Press, Washington, D.C.

National Research Council. 2002. *Effects of Trawling & Dredging on Seafloor Habitat*. National Academy Press, Washington, D.C.

Nelson, E.W., ed. 1887. Mammals. P. 267 in *Natural History Collections Made in Alaska Between the Years 1877 and 1881*, Report III, H.W. Henshaw, ed. U.S. Government Printing Office, Washington, D.C.

Niebauer, H.J. 1998. Variability in the Bering Sea ice cover as affected by a regime shift in the North Pacific in the period 1947-1996. *Journal of Geophysical Research-Oceans* 103(C12):27717-27727, 27737.

Niebauer, H.J., V. Alexander, and S.M. Henrichs. 1995. A time-series study of the spring bloom at the Bering Sea-ice edge I: Physical processes, chlorophyll and nutrient chemistry. *Continental Shelf Research* 15(15):1859-1877.

North Pacific Fishery Management Council. 1999. *Fishery Management Plan for the Groundfish Fishery in the BSAI Area*. North Pacific Fishery Management Council, Anchorage.

North Pacific Fishery Management Council. 2001a. Stock Assessment and Fishery Evaluation Report for the Groundfish Resources of the Gulf of Alaska. North Pacific Fishery Management Council, Anchorage.

North Pacific Fishery Management Council. 2001b. Stock Assessment and Fishery Evaluation Report for the Groundfish Resources of the Bering Sea/Aleutian Islands Regions. North Pacific Fishery Management Council, Anchorage.

North Pacific Fishery Management Council. 2002. Fishery Management Plan for the Groundfish of the GOA. North Pacific Fishery Management Council, Anchorage.

Northern Economics, Inc., and EDAW, Inc. 2002. Sector and regional profiles of the North Pacific groundfish fisheries—2001. North Pacific Fishery Management Council, Anchorage. Available at *http://www.fakr.noaa.gov/npfmc/Northern Economics/Northern Economics.htm*.

O'Hara, T., and T. O'Shea. 2001. Toxicology. Pp. 471-520 in *CRC Handbook of Marine Mammal Medicine*, 2nd ed., L.A. Dierauf and F.M.D. Gulland, eds. CRC Press, Boca Raton, Florida.

Ohtani, K. and T. Azumaya. 1995. Influence of interannual changes in ocean conditions on the abundance of walleye pollock (*Theragra chalcogramma*) in the eastern Bering Sea. In *Climate Change and Northern Fish Populations*. R.J. Beamish, ed. *Canadian Special Publication of Fisheries and Aquatic Sciences* 121:87-95.

Ono, K., T. Loughlin, and R. Merrick. 2001. Behavior and ecology of juvenile Steller sea lions (*Eumetopias jubatus*) during the breeding season. P. 163 in *14th Biennial Conference on the Biology of Marine Mammals*. Vancouver, Canada.

Overland, J.E., J.M. Adams, and N.A. Bond. 1999. Decadal variability of the Aleutian low and its relation to high-latitude circulation. *Journal of Climate* 12(5):1542-1548.

Paine, R.T. 1966. Food web complexity and species diversity. *The American Naturalist* 100(910):65-75.

Paine, R.T. 1980. Food webs: Linkage, interaction strength, and community infrastructure—the 3rd Tansley lecture. *Journal of Animal Ecology* 49(3):667-685.

Pascual, M.A., and M.D. Adkison. 1994. The decline of the Steller sea lion in the Northeast Pacific: Demography, harvest or environment? *Ecological Applications* 4(2):393-403.

Perez, M.A., and T.R. Loughlin. 1991. Incidental catch of marine mammals by foreign and joint venture trawl vessels in the U.S. EEZ of the North Pacific, 1973-88. NOAA Technical Report NMFS 104. U.S. Department of Commerce, National Oceanic and Atmospheric Administration, National Marine Fisheries Service, Seattle.

Peterson, W., and D. Mackas. 2001. Shifts in zooplankton abundance and species composition off central Oregon and southwestern British Columbia. *PICES Press* 9:28-31.

Pitcher, K.W. 1981. Prey of the Steller sea lion, *Eumetopias jubatus*, in the Gulf of Alaska. *Fishery Bulletin of the United States* 79(3):467-472.

Pitcher, K.W., and D.G. Calkins. 1981. Reproductive biology of Steller sea lions in the Gulf of Alaska. *Journal of Mammalogy* 62(3):599-605.

Pitcher, K.W., D.G. Calkins, and G.W. Pendelton. 1998. Reproductive performance of female Steller sea lions: An energetics-based reproductive strategy? *Canadian Journal of Zoology* 76(11):2075-2083.

Power, M.E. 1990. Effects of fish in river food webs. *Science* 250(4982):811-814.

Raum-Suryan, K.L., D.G. Calkins, and J.L. Sease. 2002. Dispersal rookery fidelity, and meta-population structure of Steller sea lions *(Eumetopias Jubatus)* in an increasing and decreasing population of Alaska. *Marine Mammal Science* 18(3):746-764.

Rea, L.D., and T. Nagy. 2000. Changes in serum leptin levels during fasting and food limitation in Steller sea lions. Pp. 171-175 in *Third Biennial Symposium of the Comparative Nutrition Society*. Pacific Grove, California.

Rea, L.D., M.A. Castellini, B.S. Fadely, and T.R. Loughlin. 1998. Health status of young Alaska Steller sea lion pups (*Eumetopias jubatus*) as indicated by blood chemistry and hematology. *Comparative Biochemistry and Physiology—Part A Molecular and Integrative Physiology* 120(4):617-623.

Rea, L.D., D.A.S. Rosen, and A.W. Trites. 2000. Metabolic responses to fasting in 6 week old Steller sea lion pups (*Eumetopias jubatus*). *Canadian Journal of Zoology* 78(5):890-894.

Rea, L.D., D.A.S. Rosen, and A.W. Trites. 2001. The effect of seasonal food limitation on adaptation to fasting in captive Steller sea lions. Abstract in 34th International Congress of Physiological Sciences. Christchurch, New Zealand.

Reed, R.K. 1995. On the variable subsurface environment of fish stocks in the Bering Sea. *Fisheries Oceanography* 4(4):317-323.

Reed, R.K., and J.D. Schumacher. 1986. Physical oceanography. Pp. 57-75 in *The Gulf of Alaska: Physical Environment and Biological Resources*, D.W. Hood and S.T. Zimmerman, eds. PB87-103230. U.S. Department of Commerce, National Oceanic and Atmospheric Administration, National Ocean Service, Ocean Assessments Division, Anchorage.

Reed, R.K., and P.J. Stabeno. 1999. The Aleutian North Slope Current. Pp. 177-191 in *Dynamics of the Bering Sea,* T.R. Loughlin and K. Ohtani, eds. AK-SG-99-03. University of Alaska Sea Grant Program, Fairbanks.

Reed, R.K., J.D. Schumacher, and L.S. Incze. 1987. Circulation in Shelikof Strait, Alaska. *Journal of Physical Oceanography* 17(9):1546-1554.

Rehberg, M.J., K.L. Raum-Suryan, K.W. Pitcher, and T.S. Gelatt. 2001. Development of juvenile Steller sea lion (*Eumetopias jubatus*) diving behaviour in Alaska. P. 177 in *14th Biennial Conference on the Biology of Marine Mammals*, Vancouver, Canada.

Reijnders, P.J.H., E.H. Ries, S. Tougaard, N. Nøergaard, G. Heidemann, J. Schwarz, E. Vareschi, and I.M. Traut. 1997. Population development of harbour seals *Phoca vitulina* in the Wadden Sea after the 1988 virus epizootic. *Journal of Sea Research* 38(1-2):161-168.

Repenning, C.A. 1976. Adaptive evolution of sea lions and walruses. *Systematic Zoology* 25(4):375-390.

Repenning, C.A., C.E. Ray, and C. Grigorescu. 1979. Pinniped biogeography. Pp. 357-369 in *Historical Biogeography, Plate Tectonics, and the Changing Environment*, J. Gray and A. Boucot, eds. Oregon State University Press, Corvallis.

Richmond, J.P., and L.D. Rea. 2001. Nutritional blood chemistry provides clues to the feeding status of young of the year and juvenile Steller sea lions (*Eumetopias jubatus*). P. 179 in *14th Biennial Conference on the Biology of Marine Mammals*. Vancouver, Canada.

Riedman, M. 1990. *The Pinnipeds: Seals, Sea Lions, and Walruses.* University of California Press, Berkeley.

Rigby, P.W., D.R. Ackley, F. Funk, H.J. Geiger, G.H. Kruse, and M.C. Murphy. 1995. Management of Marine Fisheries Resources of Alaska: A Report to the Northern Forum. Regional Information Report 5J95-04. Alaska Department of Fish & Game, Commercial Fisheries Management and Development Division, Juneau.

Rivkin, R.B., and L. Legendre. 2001. Biogenic carbon cycling in the upper ocean: Effects of microbial respiration. *Science* 291:2398-2400.

Romer, A.S. 1966. *Vertebrate Paleontology*, 3rd ed. University of Chicago Press, Chicago.

Ronholt, L.L., H.H. Shippen, and E.S. Brown. 1978. *Demersal Fish and Shellfish Resources of the Gulf of Alaska from Cape Spencer to Unimak Pass 1948-1976: A Historical Review*, Vol. 1-4. U.S. Department of Commerce, National Oceanic and Atmospheric Administration, National Marine Fisheries Service, Northwest and Alaska Fisheries Center, Seattle.

Rosen, D.A.S., and A.W. Trites. 2000. Pollock and the decline of Steller sea lions: Testing the junk-food hypothesis. *Canadian Journal of Zoology* 78(7):1243-1250.

Royer, T.C. 1998. Coastal processes in the northern North Pacific. Pp. 395-414 in *The Sea: Volume 11—The Global Coastal Ocean: Regional Studies and Synthesis*, K.H. Brink and A.R. Robinson, eds. Wiley & Sons, New York.

Royer, T.C., C.E. Grosch, and L.A. Mysak. 2001. Interdecadal variability of Northeast Pacific coastal freshwater and its implications on biological productivity. *Progress in Oceanography* 49(1-4):95-111.

Ruesink, J.L. 1998. Variation in per capita interaction strength: Thresholds due to nonlinear dynamics and nonequilibrium conditions. *Proceedings of the National Academy of Sciences of the United States of America* 95(12):6843-6847.

Saulitis, E.L., C.O. Matkin, K. Heise, L.G. Barrett Lennard, and G.M. Ellis. 2000. Foraging strategies of sympatric killer whale (*Orcinus orca*) populations in Prince William Sound, Alaska. *Marine Mammal Science* 16(1):94-109.

Scheffer, M., S. Carpenter, J.A. Foley, C. Folkes, and B. Walker. 2001. Catastrophic shifts in ecosystems. *Nature* 413:591-596.

Schoener, T.W. 1993. On the relative importance of direct versus indirect effects in ecological communities. Pp. 365-415 in *Mutualism and Community Organization*, H. Kawanabe and J.E. Cohen, eds. Oxford University Press, Oxford, U.K.

Scholin, C.A., F. Gulland, G.J. Doucette, S. Benson, M. Busman, F.P. Chavez, J. Cordaro, R. DeLong, A. DeVogalaere, J. Harvey, M. Haulena, K. Lefebvre, T. Lipscomb, S. Loscutoff, L.J. Lowenstine, R. Marin, P.E. Miller, W.A. McLellan, P.D.R. Moeller, C.L. Powell, T. Rowles, P. Silvagni, M. Silver, T. Spraker, V. Trainer, and F.M. Van Dolah. 2000. Mortality of sea lions along the central California coast linked to a toxic diatom bloom. *Nature* 403(6765):80-84.

Schumacher, J.D., and V. Alexander. 1999. Variability and role of the physical environment in the Bering Sea ecosystem. Pp. 147-160 in *Dynamics of the Bering Sea*, T.R. Loughlin and K. Ohtani, eds. AK-SG-99-03. University of Alaska Sea Grant, Fairbanks.

Schumacher, J.D., and R.K. Reed. 1986. On the Alaska coastal current in the western Gulf of Alaska. *Journal of Geophysical Research—Oceans* 91(C8):9655-9661.

Schumacher, J.D., and P.J. Stabeno. 1994. Ubiquitous eddies of the eastern Bering Sea and their coincidence with concentrations of larval pollock. *Fisheries Oceanography* 3:182-190.

Schumacher, J.D., and P.J. Stabeno. 1998. Continental shelf of the Bering Sea. Pp. 789-822 in *The Sea: Ideas and Observations of the Study of the Seas. The Global Coastal Ocean: Regional Studies and Syntheses*. Vol. 2, A.R. Robinson and K.H. Brink, eds. Wiley & Sons, New York.

Schumacher, J.D., N.A. Bond, R.D. Brodeur, P.A. Livingston, J.M. Napp, and P.J. Stabeno. In press. Climate change in the southeastern Bering Sea and some consequences for biota. In *Large Marine Ecosystems of the World—Trends in Exploitation, Protection and Research*, G. Hempel and K. Sherman, eds. Elsevier Science, Amsterdam.

Schwing, F., and C. Moore. 2000. A year without a summer for California, or a harbinger of a climate shift? *EOS, Transactions of the American Geophysical Union* 81:301, 304-305.

Sease, J.L., W.P. Taylor, T.R. Loughlin, and K.W. Pitcher. 2001. Aerial and Land-Based Surveys of Steller Sea Lions (*Eumetopias jubatus*) in Alaska, June and July 1999 and 2000. NOAA Technical Memorandum NMFS-AFSC-122. U.S. Department of Commerce, National Oceanic and Atmospheric Administration, National Marine Fisheries Service, Alaska Fisheries Science Center.

Shiganova, T.A. 1998. Invasion of the Black Sea by the ctenophore *Mnemiopsis leidyi* and recent changes in pelagic community structure. *Fisheries Oceanography* 7(3/4):305-310.

Shima, M., A.B. Hollowed, and G.R. VanBlaricom. 2002. Changes over time in the spatial distribution of walleye pollock (*Theragra chalcogramma*) in the Gulf of Alaska, 1984-1996. *Fishery Bulletin* 100(2):307-323.

Sinclair, E.H., and T.K. Zeppelin. 2002. Seasonal and spatial differences among the western stock of Steller sea lions. *Journal of Mammalogy* 83(4):973-990.

Sirenko, B.I., and V.M. Koltun. 1992. Characteristics of benthic biocenoses of the Chukchi and Bering Seas. Pp. 251-259 in *Results of the Third US-USSR Bering and Chukchi Seas Expedition (BERPAC), Summer 1988*, P.A. Nagel, ed. U.S. Fish and Wildlife Service, Washington, D.C.

Spraker, T.R., and D. Bradley. 1996. Investigations into the health status of Steller sea lions, *Eumetopias jubatus*, from 1992 to 1995. Pp. 88-108 in *Steller Sea Lion Recovery Investigations in Alaska, 1992-1994*. Alaska Department of Fish & Game, Division of Wildlife Conservation, Juneau.

Springer, A.M. 1998. Is it all climate change? Why marine bird and mammal populations fluctuate in the North Pacific. Pp. 109-119 in *Biotic Impacts of Extratropical Climate Variability in the Pacific*, G. Holloway et al., eds. Proceedings 'Aha Huliko'a Hawaiian Water Workshop, University of Hawaii, Manoa.

Springer, A.M. 2001. Still Search for a Unified Theory. Presentation to the National Research Council, Committee on the Alaska Groundfish Fishery and Steller Sea Lions, Seattle, December.

Springer, A.M., J.F. Piatt, V.P. Shuntov, G.B. Van Vliet, V.L. Vladimirov, A.E. Kuzin, and A. S. Perlov. 1999. Marine birds and mammals of the Pacific Subarctic Gyres. *Progress in Oceanography* 43(2-4):443-487.

Stabeno, P.J., and J.E. Overland. 2001. Bering Sea shifts toward an earlier spring transition. *EOS, Transactions of the American Geophysical Union* 82(29):317, 321.

Stabeno, P.J., R.K. Reed, and J.D. Schumacher. 1995. The Alaska Coastal Current: Continuity of transport and forcing. *Journal of Geophysical Research—Oceans* 100(C2):2477-2485.

Stabeno, P.J., J.D. Schumacher, and K. Ohtani. 1999. The physical oceanography of the Bering Sea. Pp. 1-28 in *Dynamics of the Bering Sea*, T.R. Loughlin and K. Ohtani, eds. AK-SG-99-03. University of Alaska Sea Grant Program, Fairbanks.

Stabeno, P.J., N.A. Bond, N.B. Kachel, S.A. Salo, and J.D. Schumacher. 2001. On the temporal variability in the physical environment over the south-eastern Bering Sea. *Fisheries Oceanography* 10(1):81-98.

Stabeno, P.J., N.A. Bond, A.J. Hermann, C.W. Mordy, and J.E. Overland. In press. Meteorology and oceanography of the northern Gulf of Alaska. *Progress in Oceanography*.

Stokesbury, K.D.E., J. Kirsch, E.V. Patrick, and B.L. Norcross. 2002. Natural mortality estimates of juvenile Pacific herring (*Clupea pallasi*) in Prince William Sound, Alaska. *Canadian Journal of Fisheries and Aquatic Sciences* 59(3):416-423.

Sugimoto, T., and K. Tadokoro. 1997. Interannual-interdecadal variations in zooplankton biomass, chlorophyll concentration and physical environment in the subarctic Pacific and Bering Sea. *Fisheries Oceanography* 6(2):74-93.

Swartzman, G., W. Stuetzle, K. Kulman, and M. Powojowski. 1994. Relating the distribution of pollock schools in the Bering Sea to environmental factors. *ICES Journal of Marine Science* 51(4):481-492.

Thomas, G.L., and R.E Thorne. 2001. Night-time predation by Steller sea lions. *Nature* 411:1013.

Thompson, G.G., and M.W. Dorn. 2001. Assessment of the Pacific cod stock in the eastern Bering Sea and Aleutian Islands areas. Pp. 2-1–2-75 in *Stock Assessment and Fishery Evaluation Report for the Groundfish Resources of the Bering Sea/Aleutian Islands Regions*, J. Blackburn et al., eds. North Pacific Fishery Management Council, Anchorage.

Thompson, G.G., H.H. Zenger, and M.W. Dorn. 2001. Assessment of the Pacific cod stock in the Gulf of Alaska. Pp. 2-1–2-73 in *Stock Assessment and Fishery Evaluation Report for the Groundfish Resources of the Gulf of Alaska*, J. Blackburn et al., eds. North Pacific Fishery Management Council, Anchorage.

Thompson, W.F., F.V. Thorsteinson, and D.E. Bevan. 1955. Present status of sea lion investigations in the Alaska Peninsula and Kodiak Island areas. *Circular* 75 (Peninsula Memorandum 4), Seattle.

Thorsteinson, F.V., and C.J. Lensink. 1962. Biological observations of Steller sea lions taken during an experimental harvest. *Journal of Wildlife Management* 26(4):353-359.

Thorsteinson, F.V., R.W. Nelson, and D.F. Lall. 1961. Experimental harvest of the Steller sea lion in Alaska waters. Special Scientific Report Fisheries No. 371. U.S. Department of Interior, Fish and Wildlife Service, Bureau of Commercial Fisheries, Washington, D.C.

Tilman, D., and P. Kareiva, eds. 1997. *Spatial Ecology: The Role of Space in Population Dynamics and Interspecific Interactions*. Princeton University Press, Princeton.

Trenberth, K.E. 1990. Recent observed interdecadal climate changes in the northern hemisphere. *Bulletin of the American Meteorological Society* 71(7):988-993.

Trenberth, K.E., and J.W. Hurrell. 1994. Decadal atmosphere-ocean variations in the Pacific. *Climate Dynamics* 9:303-319.

Trites, A.W. 2001. North Pacific Universities Marine Mammal Research Consortium: Recent Findings. Presentation to the National Research Council, Committee on the Alaska Groundfish Fishery and Steller Sea Lions, Seattle, December.

Trites, A.W., and P.A. Larkin. 1992. The Status of Steller Sea Lion Populations and the Development of Fisheries in the Gulf of Alaska and Aleutian Islands. Report of the Pacific States Marine Fisheries Commission pursuant to National Oceanic and Atmospheric Administration Award No. NA17FD0177. Fisheries Centre, University of British Columbia, Vancouver.

Trites, A.W., and P.A. Larkin. 1996. Changes in the abundance of Steller sea lions (*Eumetopias jubatus*) in Alaska from 1956 to 1992: How many were there? *Aquatic Mammals* 22:153-166.

Trites, A.W., P.A. Livingston, S. Mackinson, M.C. Vasconcellos, A.M. Springer, and D. Pauly. 1999. Ecosystem Change and the Decline of Marine Mammals in the Eastern Bering Sea: Testing the Ecosystem Shift and Commercial Whaling Hypotheses. Fisheries Centre Research Reports Vol. 7, No. 1. University of British Columbia, Vancouver.

Vandenbosch, R. 2000. Effects of ENSO and PDO events on seabird populations as revealed by Christmas Bird Count data. *Waterbirds* 23(3):416-422.

Vania, J.S. 1972. *Marine Mammals Investigation: Sea Lion Pelage Study. Federal Aid to Wildlife Restoration*. Final Report, Projects W-17-3 and W-17-4. Alaska Department of Fish & Game, Juneau.

Waite, J.M., N. Friday, and S.E. Moore. 2002. Killer whale (*Orcinus orca*) distribution and abundance for central and southeastern Bering Sea, July 1999 and June 2000. *Marine Mammal Science* 18(4):779.

Walsh, J.J. 1995. DOC storage in Arctic seas: The role of continental shelves. Pp. 203-230 in *Arctic Oceanography: Marginal Ice Zones and Continental Shelves*, J.M. Grebmeier and W.O. Smith, eds. American Geophysical Union, Washington, D.C.

Walsh, J.J., C.P. McRoy, L.K. Coachman, J.J. Goering, J.J. Nihoul, T.E. Whitledge, T.H. Blackburn, P.L. Parker, C.D. Wirick, P.G. Shuert, J.M. Grebmeier, A.M. Springer, R.D. Tripp, D.A. Hansell, S. Djenidi, E. Deleersnijder, K. Henricksen, B.A. Lund, P. Andersen, F.E. Müller-Karger, and K. Dean. 1989. Carbon and nitrogen cycling within the Bering/Chukchi Seas: Source regions for organic matter effecting AOU demands of the Arctic Ocean. *Progress in Oceanography* 22(4):277-359.

Walters, C.J. 1986. *Adaptive Management of Renewable Resources*. Macmillan, New York.

Walters, C., and J.F. Kitchell. 2001. Cultivation/depensation effects on juvenile survival and recruitment: Implications for the theory of fishing. *Canadian Journal of Fisheries and Aquatic Sciences* 58(1):39-50.

Walters, C.J., V. Christensen, and D. Pauly. 1997. Structuring dynamic models of exploited ecosytems from trophic mass-balance assessments. *Reviews in Fish Biology and Fisheries* 7(2):139-172.

Walters, C.J., V. Christensen, D. Pauly, and J.F. Kitchell. 2000. Representing density dependent consequences of life history strategies in aquatic ecosystems: EcoSim II. *Ecosystems* 3(1):70-83.

Ware, D.M. 1995. A century and a half of change in the climate of the NE Pacific. *Fisheries Oceanography* 4(4):267-277.

Wespestad, V.G. 1991. Pacific herring population dynamics, early life history, and recruitment variation relative to eastern Bering Sea oceanographic factors. Ph.D. dissertation, University of Washington, Seattle.

Wespestad, V.G., and D.R. Gunderson. 1991. Climate induced variation in eastern Bering Sea herring recruitment. Pp. 127-140 in *Proceedings of the International Herring Symposium, Anchorage, Alaska, October 23-25, 1990*. Alaska Sea Grant College Program Report 91-01. University of Alaska, Fairbanks.

Wilderbuer, T.K., and T.M. Sample. 2001. Arrowtooth flounder. Pp. 5-1–5-12 in *Stock Assessment and Fishery Evaluation Report for the Groundfish Resources of the Bering Sea/Aleutian Islands*. North Pacific Fisheries Management Council, Anchorage.

Wilke, F., and K.W. Kenyon. 1952. Notes on the food of fur seal, sea-lion, and harbor porpoise. *Journal of Wildlife Management* 16(3):396-397.

Wilson, C., A. Hollowed, M. Shima, P. Walline, and S. Stienessen. 2002. Interactions between commercial fishing and walleye pollock. Pp. 1-14 in *AFSC Quarterly Report* (January-February-March 2002). U.S. Department of Commerce, National Oceanic and Atmospheric Administration, National Marine Fisheries Service, Alaska Fisheries Science Center, Seattle.

Wilson, J.G., and J.E. Overland. 1986. Meteorology. Pp. 31-54 in *The Gulf of Alaska: Physical Environment and Biological Resources*, D.W. Hood and S.T. Zimmerman, eds. PB87-103230. U.S. Department of Commerce, National Oceanic and Atmospheric Administration, National Ocean Service, Ocean Assessments Division, Anchorage.

Witherell, D., N. Kimball, and J. DiCosimo. 2000. *Status and Trends of Principal Groundfish and Shellfish Stocks in the Alaska EEZ, 2001*. North Pacific Fishery Management Council, Anchorage.

Wolfe, R.J., and L.B. Hutchinson-Scarbrough. 1999. The subsistence harvest of harbor seal and sea lions by Alaska natives in 1998. Technical Paper No. 250. Alaska Department of Fish & Game, Juneau.

Wolfe, R.J., and C. Mishler. 1997. The subsistence harvest of harbor seal and sea lion by Alaska natives in 1996. Technical Paper No. 241. Alaska Department of Fish & Game, Juneau.

Wooster, W.S., and A.B. Hollowed. 1995. Decadal-scale variations in the eastern subarctic Pacific. I. Winter ocean conditions. Pp. 81-85 in *Climate Change and Northern Fish Populations*, R.J. Beamish, ed. *Canadian Special Publication-Fisheries and Aquatic Science* 121.

Wootton, J.T. 1994. The nature and consequences of indirect effects in ecological communities. *Annual Review of Ecology and Systematics* 25:443-466.

Wyllie-Echevarria, T., and W.S. Wooster. 1998. Year-to-year variations in Bering Sea ice cover and some consequences for fish distributions. *Fisheries Oceanography* 7(2):159-170.

Wynen, L.P., S.D. Goldsworthy, S.J. Insley, M. Adams, J.W. Bickham, J. Francis, J.P. Gallo, A.R. Hoelzel, P. Majluf, R.W.G. White, and R. Slade. 2001. Phylogenetic relationships within the eared seals (*Otariidae: Carnivora*): Implications for the historical biogeography of the family. *Molecular Phylogenetics and Evolution* 21(2):270-284.

York, A.E. 1994. The population dynamics of northern sea lions, 1975-1985. *Marine Mammal Science* 10(1):38-51.

York, A.E., R.L. Merrick, and T.R. Loughlin. 1996. An analysis of the Steller sea lion metapopulation in Alaska. Pp. 259-292 in *Metapopulations and Wildlife Conservation*, D.R. McCullough, ed. Island Press, Washington, D.C.

Zarnke, R.L., T.C. Harder, H.W. Vos, J.M. ver Hoef, and A.D.M.E. Osterhaus. 1997. Serologic survey for phocid herpesvirus-1 and -2 in marine mammals from Alaska and Russia. *Journal of Wildlife Diseases* 33(3):459-465.

Zenteno-Savin, T., M.A. Castellini, L.D. Rea, and B.S. Fadely. 1997. Plasma haptoglobin levels in threatened Alaskan pinniped populations. *Journal of Wildlife Diseases* 33(1):64-71.

Zheng, J., and G.H. Kruse. 2000. Recruitment patterns of Alaskan crabs in relation to decadal shifts in climate and physical oceanography. *ICES Journal of Marine Science* 57(2):438-451.

# Appendixes

# APPENDIX A

# Committee and Staff Biographies

## COMMITTEE MEMBERS

**Robert T. Paine** (*Chair*) earned his Ph.D. from the University of Michigan in 1961. He recently retired (1998) from his position as professor and former chairman of the Zoology Department at the University of Washington, where he worked since 1962. His research interests include experimental ecology of organisms on rocky shores, interrelationships between species in an ecosystem, and the organization and structure of marine communities. He has examined the roles of predation and disturbance in promoting coexistence and biodiversity. Dr. Paine is a member of the National Academy of Sciences and was a member of the National Research Council's Board of Life Sciences. He is currently serving on two Academy committees: Committee on the Protection of Ecology and Resources of the Caspian Sea and the Temporary Nominating Group for Global and Human Environmental Sciences. His extensive National Research Council experience includes committee service for the reports *Understanding Marine Biodiversity* and *Sustaining Marine Fisheries*.

**Daniel W. Bromley** earned his Ph.D. in natural resource economics from Oregon State University in 1969. Dr. Bromley is currently the Anderson-Bascom Professor of Applied Economics at the University of Wisconsin, Madison. His research interests include institutional economics, political economy, natural resource economics, and environmental implications of economic development. He served as president of the International Association for the Study of Common Property. Dr. Bromley was chair of the National Research Council's Committee on Common Property Resource Management and served as a member of the National Research Council's Committee to Review the Community Development Quota Program in Alaska.

**Michael A. Castellini** earned his Ph.D. in marine biology from the Scripps Institution of Oceanography in 1981. Dr. Castellini is director of the Institute of Marine Science at the University of Alaska, Fairbanks. His research interests focus on adaptations of marine mammals to life in the sea, including diving physiology, nutritional biochemistry, behavior, development, and survival in species such as Steller sea lions, harbor seals, Weddell seals, and elephant seals. His laboratory is also working on lipid metabolism in marine mammals, the biochemistry of contaminants, metal and antioxidant chemistry, and immune function. He is a charter member of the Society for Marine Mammalogy, past science director for the Alaska SeaLife Center in Seward, and a member and past officer of the American Physiological Society.

**Larry B. Crowder** earned his Ph.D. in zoology from Michigan State University, East Lansing, in 1978. Dr. Crowder is currently the Stephen Toth Professor in Marine Ecology at Duke University. His research interests include species interactions and community ecology of aquatic systems, including freshwater and marine systems, predation, competition, and food web interactions. He has studied fisheries recruitment and population dynamics. He also has been involved in population modeling and data analysis to address various management scenarios for threatened and endangered species, especially those taken as bycatch. He has served on the editorial board for *Ecology* and *Ecological Applications*. He is a member of the International Union for Conservation of Nature and Natural Resources Marine Turtle Specialist Group and serves on the National Research Council's Ocean Studies Board. Dr. Crowder was a member of the Program Management Committee for the South Atlantic Bight Recruitment Experiment.

**James A. Estes** earned his Ph.D. in biology and statistics from the University of Arizona in 1974. Dr. Estes is a research biologist with the Western Ecological Research Center of the U.S. Geological Survey in Santa Cruz, California, where he has worked for the past 20 years. He is also a research associate and adjunct professor with the Center for Marine Studies at the University of California, Santa Cruz. His research interests include wildlife ecology; predation and conservation; and more specifically killer whale interactions with sea otters and Steller sea lions; and spatial, temporal, and functional dimensions of sea otter and kelp forest interactions. He has served on the Board of Editors for the Ecological Society of America, as deputy chairman for the Species Survival Commission of the International Union for Conservation of Nature and Natural Resources, and as cochairman for the Standing Committee on Marine Mammals of the American Society of Mammalogists.

**Jacqueline M. Grebmeier** earned her Ph.D. in biological oceanography from the University of Alaska, Fairbanks, in 1987. Dr. Grebmeier is currently a research associate professor and project director at the University of Tennessee, Department of Ecology and Evolutionary Biology. Her research interests include pelagic-benthic coupling, benthic carbon cycling, and benthic faunal population structures in the marine environment in both the Arctic and the Antarctic. Dr. Grebmeier has served as a member of the U.S. Arctic Research Commission since 2000. She is involved with research efforts for the National Science Foundation's global change research project on shelf-basin interactions and ecosystem change studies in the Bering and Chukchi Seas. She served on the National Research Council's committee for the report *The Bering Sea Ecosystem.*

**Frances M.D. Gulland** earned her Ph.D. and doctorate in veterinary medicine from Cambridge University, Great Britain. She is currently the director of veterinary science at the Marine Mammal Center in Sausalito, California. Her research interests include sources of mortality in marine mammal populations, and she was the first to identify domoic acid from the phytoplankton *Pseudonitzschia australis* as the cause of death for several hundred California sea lions in the Monterey Bay region in 1998. She is also a member of the Committee of Scientific Advisors of the U.S. Marine Mammal Commission.

**Gordon H. Kruse** earned his Ph.D. in fisheries science in 1983 from Oregon State University. Dr. Kruse worked as a marine fisheries scientist at the Alaska Department of Fish & Game from 1989 until 2001. In November 2001, Dr. Kruse began a new position as the president's professor of fisheries at the University of Alaska, Fairbanks, Juneau Center, School of Fisheries and Ocean Sciences. His research focuses on population estimation models, stock production parameters, population dynamics, alternative management strategies, fishery oceanography, and ecosystem dynamics. Dr. Kruse is a delegate to the Fishery Science Committee of the North Pacific Marine Science Organization, and he chaired the state of Alaska's Steller Sea Lion Restoration Team. He recently served on the National Research Council's Committee on Ecosystem Effects of Fishing: Phase 1—Effects of Bottom Trawling on Seafloor Habitats.

**Nathan J. Mantua** earned his Ph.D. in atmospheric sciences from the University of Washington in 1994. His graduate studies focused on the dynamics of El Niño and the Southern Oscillation. For the past 6 years he has worked as a research scientist and affiliate assistant professor in Atmospheric Sciences and Marine Affairs at the University of Washington. In April 2000 he received the National Oceanic and Atmospheric Administration's Presidential Early Career Award for Scientists and Engineers for his research on climate impacts on ecosystems and society. He currently serves as a scientific advisor to the U.S. delegation of the North Pacific Anadromous Fisheries Commission and is a member of the scientific steering committee for the U.S. Global Oceans-Ecosystems Dynamics research program.

**James D. Schumacher (a.k.a. Two Crow)** earned his Ph.D. in marine sciences from the University of North Carolina, Chapel Hill, in 1974. He was an oceanographer at the National Oceanic and Atmospheric Administration's (NOAA's) Pacific Marine Environmental Laboratory until 1998 and served as director of NOAA's Fisheries-Oceanography Coordinated Investigations. Since 1998 he has been president of Two Crow Environmental, Inc. His research is on pathways that link changes in physical phenomena to fluctuations in biological populations and the combined impact of physical and biological changes on native peoples. He is currently involved in the Southeast Bering Sea Carrying Capacity, a NOAA Coastal Ocean Program, where he is examining indices of the environment that may contribute to fluctuations in pollock populations.

**Donald B. Siniff** earned his Ph.D. in entomology, fisheries, and wildlife from the University of Minnesota in 1967. Dr. Siniff is currently a professor in the Department of Ecology, Evolution, and Behavior at the University of Minnesota, St. Paul. His research interests include population biology of marine mammals. He is a past member of the Marine Mammal Commission and the National Research Council's Polar Research Board. He is a member of the U.S. Fish and Wildlife Service's Southern Sea Otter and the National Marine Fisheries Service's Northern Sea Lion Recovery Teams.

**Carl J. Walters** earned his Ph.D. in fisheries from Colorado State University in 1969. He is a professor at the Fisheries Centre at the University of British Columbia, with a joint appointment in zoology. His areas of research include the dynamics of ecological communities, application of mathematical models and computer simulation techniques to problems in resource ecology, and adaptive management of renewable resources. He is a fellow of the Royal Society of Canada. He served on the National Research Council's committee for the report *Improving Fish Stock Assessment*.

## NATIONAL RESEARCH COUNCIL STAFF

**Susan Roberts** (*Study Director*) received her Ph.D. in marine biology from the Scripps Institution of Oceanography. Dr. Roberts is a senior program officer for the National Research Council's Ocean Studies Board, where she has directed a number of studies, including the following reports: *Effects of Trawling & Dredging on Seafloor Habitat* (2002), *Marine Protected Areas—Tools for Sustaining Ocean Ecosystems* (2001); *Bridging Boundaries Through Regional Marine Research* (2000), and *From Monsoons to Microbes—Understanding the Ocean's Role in Human Health* (1999). Dr. Roberts' research interests include marine microbiology, fish physiology, and marine biotechnology.

**Nancy A. Caputo** received a master's of public policy from the University of Southern California and a bachelor's degree in political science/international relations. During her tenure with the Ocean Studies Board, she has assisted with the completion of two reports: *A Review of the Florida Keys Carrying Capacity Study* (2002) and *Emulsified Fuels—Risks and Response* (2002). Ms. Caputo has previous professional experience researching fisheries management in the northeastern and northwestern United States, socioeconomic assistance programs for fishing communities, and habitat restoration programs. Her interests include marine mammal policy and science, ocean policy, coastal management, and habitat restoration.

# Appendix B

# Acronyms

| | |
|---|---|
| ABC | acceptable biological catch |
| ACC | Alaskan Coastal Current |
| ADF&G | Alaska Department of Fish & Game |
| AFA | American Fisheries Act |
| ASLC | Alaska SeaLife Center |
| ASSLRT | Alaska Steller Sea Lion Recovery Team |
| BiOp #1 | Biological Opinion #1, released December 1998 |
| BiOp #2 | Biological Opinion #2, released January 1999 |
| BiOp #3 | Biological Opinion #3, released November 2000 |
| BiOp #4 | Biological Opinion #4, released June 2001 |
| BSAI | Bering Sea/Aleutian Islands |
| CPUE | catch per unit effort |
| DDT | dichloro-diphenyl-trichloro-ethane |
| DMV | dolphin morbillivirus |
| EEZ | Exclusive Economic Zone |
| EIS | environmental impact statement |
| ENSO | El Niño-Southern Oscillation |
| ESA | Endangered Species Act of 1973 |
| GOA | Gulf of Alaska |

| | |
|---|---|
| MSY | maximum sustainable yield |
| mt | metric tons |
| mtDNA | mitochondrial DNA |
| | |
| nm | nautical mile |
| NMFS | National Marine Fisheries Service |
| NOAA | National Oceanic and Atmospheric Administration |
| NPFMC | North Pacific Fishery Management Council |
| NPPI | North Pacific Pressure Index |
| | |
| OCH | oscillating control hypothesis |
| | |
| PCB | polychlorinated biphenyl |
| PDO | Pacific Decadal Oscillation |
| PDV | phocine distemper virus |
| PWS | Prince William Sound |
| | |
| RPAs | reasonable and prudent alternatives |
| | |
| SEAK | Southeast Alaska |
| SLP | sea level pressure |
| SSB | spawning stock biomass |
| SSL | Steller sea lion |
| SST | sea surface temperature |
| | |
| TAC | total allowable catch |

# Appendix C

# Glossary

**Acceptable biological catch (ABC)**—a preliminary description of acceptable harvest (or range of harvests) for a given stock or stock complex. Its derivation focuses on the status and dynamics of the stock, environmental conditions, other ecological factors, and prevailing technological characteristics of the fishery. The fishing mortality rate used to calculate ABC is constrained to be lower than rates associated with overfishing.

**Catch per unit effort (CPUE)**—an index showing the ratio of a catch of fish, in numbers or in weight, and a standard measure of the fishing effort expended to catch them.

**Maximum sustainable yield (MSY)**—the largest average catch or yield that can be continuously taken from a stock under existing environmental conditions.

**Metapopulation**—a set of local populations in some larger area, where migration between patches is possible (Hanski and Simberloff, 1997).

**Odobenids**—one of the three families of pinnipeds. This family is primarily made up of walruses.

**Otariids**—one of the three families of pinnipeds. This family is primarily made up of sea lions and fur seals.

**Phocids**—one of the three families of pinnipeds. This family is primarily made up of true seals.

**Recruitment overfishing**—the rate of fishing above which recruitment to the exploitable stock becomes significantly reduced. This is characterized by a greatly reduced spawning stock, a decreasing proportion of older fish in the catch, and generally very low recruitment year after year.

**Spawning stock biomass (SSB)**—the total weight of all sexually mature fish in the population. This quantity depends on year-class abundance, the exploitation pattern, the rate of growth, fishing and natural mortality rates, the onset of sexual maturity, and environmental conditions.

**Total allowable catch (TAC)**—an annually determined catch that is species specific and based on consideration of maximum sustainable yield, equilibrium yield, and optimum yield for the groundfish complex as a whole.

# APPENDIX D

# Early Account of Steller Sea Lions

SOURCE: Nelson, E.W., ed. 1887. Mammals. P. 267 in *Natural History Collections Made in Alaska Between the Years 1877 and 1881*, Report III, H.W. Henshaw, ed. U.S. Government Printing Office, Washington, D.C.

# REPORT

UPON

# NATURAL HISTORY COLLECTIONS

MADE IN

# ALASKA

BETWEEN THE YEARS 1877 AND 1881

EDWARD W. NELSON.

EDITED BY HENRY W. HENSHAW

PREPARED UNDER THE DIRECTION OF THE CHIEF SIGNAL OFFICER

---

No. III.

ARCTIC SERIES OF PUBLICATIONS ISSUED IN CONNECTION WITH THE SIGNAL SERVICE, U.S. ARMY.

WITH 31 PLATES.

---

WASHINGTON:
GOVERNMENT PRINTING OFFICE.
1887.

Formerly they were abundant all along the Aleutian chain. They are now so scarce among these islands, and the ones that are found there frequent places so difficult of access, that the Aleuts secure very few of them each year. They are still rather common at a few points along the north shore of Unimak Island and the peninsula of Alaska, while small parties are found scattered all along the Aleutian chain, hauling up on certain rocky points and shelves facing the sea, most of which are well known localities to the Aleuts.

In May, 1877, I saw a small party on the rocks on the north shore of Akoutan, and during the same month a fierce storm outside brought a few of them into the harbor at Unalaska. North of the Fur Seal Islands they are extremely rare or unknown at present, although I learned from the Eskimo of their occasional occurrence north to the Yukon mouth and about the shore of Nunevak Island. Prom the Aleutian Islands eastward and southward they occur all along the coast to California, where their range overlaps that of the southern species.

Large males of Steller's Sea-lion are from 11 to 18 feet long, according to Mr. Elliott, and weigh about a thousand pounds. The females are much smaller, and weigh about four or five hundred pounds.

After the annual catch of fur seals is secured on the Seal Islands, a drive of several hundred sea-lions is made to procure the skins used in covering the large native boats or umiaks. A few years ago this drive was made very easily, and an abundance of animals found, but at present they are becoming much fewer, and it is almost or quite impossible to secure the full number. It is probably a matter of but a few years before they mill become rare or unknown upon these islands, where they were formerly more numerous than anywhere else.

Like the fur seal, this animal is migratory, arriving at its breeding grounds on the Fur Seal Islands in May, and the last of them leave there when the severe winter weather begins, about the first of January. Their migration is not so general as that, of the fur seal, as some of them are found about the Seal Islands the entire winter during mild seasons.

Mr. Elliott claims that the flesh of a young sea-lion is tender, juicy, and something like veal, but becomes rank and tough when the animal approaches maturity. The same may be said of the flesh of the fur seal. The first of the latter meat I ever ate was at Unalaska, and as there was a flock of sheep there at the time I was entirely deceived, thinking I had been eating mutton until told that it was young fur seal. The meat had the color and flavor of good mutton.

The natives of the Seal Islands claim that nearly seventy years ago the sea-lions alone occupied nearly all of the shore line of Saint George Island, and numbered several hundred thousand individuals. By direction of the Russians they were driven off repeatedly until they left the place, and the shore was then occupied by fur seals.

These northern sea lions have a "deep base growl and a prolonged, steady roar," quite unlike the barking note so characteristic of the southern sea-lion of the California coast. To the natives of the Fur Seal and Aleutian Islands this animal is of the same value as the walrus is to the Eskimo of the coast to the northward. Its skin, flesh, intestines, bones, sinews, and oil all come into play as food or in the simple manufactures of the Aleuts.

Like the fur seal they have a dreaded enemy in the Killer Whale, which pursues and captures them at sea and about their rocky resorts. The native hunters when at sea frequently see them leaping high out of the water in useless endeavor to escape their pursuers. At such times they say it is dangerous for an umiak or other small boat to be in the vicinity, as the animal, in its terror, will sometimes leap into and wreck the boat. They are hunted with gun and spear in the Aleutian Islands, but, like most seals, if shot in the water in summer they will sink at once, owing to the small amount of fat on them at that season.

In common with the fur seal, this species has the habit of swallowing stones. Mr. Elliott found stones weighing a pound or two in their stomachs, and preserved one stomach containing over 10 pounds of such stones.

In the North the young are brought forth in June.

# Appendix E

# Federal Funding Summary

Federal Funding ($1,000s) for Steller Sea Lion Research in Fiscal Years (FY) 1982–2002 Available to the National Marine Fisheries Service and Other Federal Agencies,[a] and to Other Research Organzations.[b]

| FY | Federal Agencies | Nonfederal Research Organizations | Total |
|---|---|---|---|
| 1982 | 50 | | 50 |
| 1983 | 50 | | 50 |
| 1984 | 50 | | 50 |
| 1985 | 200 | | 200 |
| 1986 | 100 | | 100 |
| 1987 | 300 | | 300 |
| 1988 | 200 | | 200 |
| 1989 | 200 | | 200 |
| 1990 | 600 | | 600 |
| 1991 | 559 | | 559 |
| 1992 | 676 | 750 | 1,426 |
| 1993 | 517 | 728 | 1,245 |
| 1994 | 584 | 708 | 1,292 |
| 1995 | 645 | 708 | 1,353 |
| 1996 | 831 | 1,005 | 1,836 |
| 1997 | 867 | 1,022 | 1,889 |
| 1998 | 978 | 2,043 | 3,021 |
| 1999 | 1,653 | 1,863 | 3,516 |
| 2000 | 2,110 | 2,610 | 4,720 |
| 2001 | 15,850 | 27,300[c] | 43,150 |
| 2002 | 25,650 | 14,495 | 40,145 |

[a]National Oceanic and Atmospheric Administration's Office of Oceanic and Atmospheric Research and National Ocean Service.

[b]Alaska SeaLife Center, Alaska Department of Fish and Game, North Pacific Universities Marine Mammal Research Consortium, North Pacific Fishery Management Council, University of Alaska, and Alaska Fisheries Development Foundation.

[c]Includes funds for the Steller Sea Lion Research Initiative.

SOURCE: Ferrero and Fritz (2002, Table 1).

# Appendix F

# Meeting Agendas

**MEETING 1**

**Doubletree Hotel**
**Seattle, Washington**
**August 22-23, 2001**

**WEDNESDAY, AUGUST 22, 2001**

*OPEN SESSION*

| | |
|---|---|
| 10:30 a.m. | **Welcome and Introductions**—Bob Paine, *Chair*, and Susan Roberts, *Study Director* |
| 11:00 | **Presentations from the Sponsor**—Chris Oliver and David Witherell, *North Pacific Fishery Management Council* |
| 12:00 p.m. | **Discussion** |
| 12:30 | **Lunch** |
| 1:00 | **Presentation from the National Marine Fisheries Service**—Doug DeMaster, *National Marine Mammal Laboratory, Alaska Fisheries Science Center* |
| 3:00 | **Discussion** |
| 3:30 | **Break** |
| 4:00 | **Presentation**—Jack Tagart, *Washington Department of Fish and Wildlife, Olympia and Vice Chair, Science and Statistical Committee, North Pacific Fishery Management Council* |

| | |
|---|---|
| 4:30 | **Discussion** |
| 5:00 | **Open session adjourns for the day** |

## THURSDAY, AUGUST 23, 2001

*OPEN SESSION*

| | |
|---|---|
| 8:00 a.m. | **Breakfast** |
| 8:45 | **Welcome**—Bob Paine, *Chair*, and Susan Roberts, *Study Director* |
| 9:00 | **Review of the Alaska fisheries and Steller sea lions**—Gordon Kruse, *Chair, Alaska Steller Sea Lion Restoration Team* |
| 10:30 | **Break** |
| 10:45 | **Discussion** |
| 12:00 p.m. | **Meeting adjourns** |

# MEETING 2

**Hotel Captain Cook**
**Anchorage, Alaska**
**October 29-30, 2001**

## MONDAY, OCTOBER 29, 2001

*OPEN SESSION*

| | |
|---|---|
| 8:00 a.m. | **Breakfast** |
| 8:15 | **Welcome**—Bob Paine, *Chair*, and Susan Roberts, *Study Director* |
| 8:30 | **Dave Fraser, Owner and Captain, *F/V Muir Milach*** |
| 9:00 | **Karl Haflinger, Sea State, Inc.** |
| 9:30 | **Discussion** |
| 9:45 | **Break** |
| 10:00 | **Oliver Holm, Kodiak** |
| 10:30 | **Helen Chythlook, Bristol Bay Native Association** |
| 11:00 | **Frank Logusak, Togiak Traditional Council** |
| 11:30 | **Discussion** |
| 12:00 p.m. | **Lunch** |
| 1:00 | **Jack Sterne, Trustees for Alaska**<br>*An Overview of the Steller Sea Lion Litigation and Its Relationship to the National Research Council's Committee on Alaska Groundfish Fishery and Steller Sea Lions* |
| 1:30 | **Ken Stump, consultant for Greenpeace and colitigants** |

| | |
|---|---|
| 2:00 | **David Cline, World Wildlife Fund**<br>*World Wildlife Fund Perspective on Steller Sea Lion Issues* |
| 2:30 | **Linda Larson, Attorney for United Catcher Boats in *Greenpeace v. National Marine Fisheries Service*** |
| 3:00 | **Break** |
| 3:30 | **Gordon Kruse, member of the Committee on the Alaska Groundfish Fishery and Steller Sea Lions**<br>*Videotaped interview with Ed Opheim, Kodiak—an historical perspective on Steller sea lions and the Alaskan fisheries* |
| 4:00 | **Discussion and public comment** |
| 5:30 | **Meeting adjourns for the day** |

## TUESDAY, OCTOBER 30, 2001

*OPEN SESSION*

| | |
|---|---|
| 8:00 a.m. | **Breakfast** |
| 8:15 | **Welcome and Introductions**—Bob Paine, *Chair*, and Susan Roberts, *Study Director* |
| 8:30 | **Shane Capron, National Marine Fisheries Service, Juneau**<br>*Summary of the August 2001 Biological Opinion* |
| 9:00 | **Dan Goodman, Montana State University**<br>*Findings from the Independent Reviews of Past Biological Opinions* |
| 9:30 | **Larry Cotter, Aleutian Pribilof Island Community Development Association (Chair of the RPA committee)**<br>*Update on the Reasonable and Prudent Alternatives* |
| 10:00 | **Discussion** |
| 10:15 | **Break** |
| 10:30 | **Vladimir Burkanov, Natural Resources Consultants, Inc.**<br>*Steller Sea Lion Populations in Japan and Russia* |
| 11:00 | **Ken Pitcher, Alaska Department of Fish & Game** |
| 11:30 | **Fritz Funk, Alaska Department of Fish & Game**<br>*The Herring Fishery in Alaska* |
| 12:00 p.m. | **Discussion** |
| 12:15 | **Open session adjourns** |

## MEETING 3

**The University Inn**
**4140 Roosevelt Way, NE**
**Seattle, WA 98105**
**December 9-12, 2001**

### MONDAY, DECEMBER 10, 2001

*OPEN SESSION*

| | |
|---|---|
| 8:00 a.m. | **Breakfast** |
| 8:15 | **Welcome**—Bob Paine, *Chair*, and Susan Roberts, *Study Director* |
| 8:30 | **Kerim Aydin,** University of Washington<br>*Ecopath and Ecosim in the Eastern Bering Sea: An Evaluation Focusing on Steller Sea Lion Issues* |
| 9:00 | **Andrew Trites,** Research Director, North Pacific Universities Marine Mammal Research Consortium, University of British Columbia<br>*North Pacific Universities Marine Mammal Research Consortium: Recent Research Findings* |
| 9:30 | **Discussion** |
| 9:45 | **Break** |
| 10:00 | **Herb Maschner,** Department of Anthropology, Idaho State University |
| 10:30 | **Boris Worm,** Biology Department, Dalhousie University, Halifax, Nova Scotia<br>*Cascading Fishery Effects and Multiple Stable States in Ocean Food Webs* |
| 11:00 | **Doug DeMaster**, Director, Alaska Fisheries Science Center, National Marine Fisheries Service, National Oceanic and Atmospheric Administration<br>*Population Trends in Several Marine Mammal Populations in the North Pacific* |
| 11:30 | **Discussion** |
| 12:00 p.m. | **Lunch** |
| 1:00 | **Alan Springer,** Institute of Marine Science, University of Alaska, Fairbanks |
| 1:30 | **George Hunt,** Department of Ecology and Evolutionary Biology, University of California, Irvine |
| 2:00 | **Vernon Byrd,** U.S. Fish and Wildlife Service, Homer, Alaska<br>*Patterns of Change in Populations of Seabirds in the Western Gulf of Alaska and Southern Bering Sea Regions Since the Mid-1970s* |
| 2:30 | **Discussion** |

| | |
|---|---|
| 2:45 | **Break** |
| 3:00 | **Doug Eggers,** Alaska Department of Fish & Game, Juneau<br>*An Overview of the Spatial and Temporal Distribution of Fisheries for Herring, Salmon, and Pacific Cod, Managed by the State of Alaska* |
| 3:30 | **Lowell Fritz,** Alaska Fisheries Science Center, National Marine Fisheries Service, National Oceanic and Atmospheric Administration<br>*North Pacific Groundfish, Groundfish Fisheries, and Steller Sea Lions* |
| 4:00 | **Glenn VanBlaricom,** University of Washington<br>*Viability Models for the Conservation Status of Steller Sea Lions* |
| 4:30 | **Discussion** |
| 5:30 | **Meeting adjourns for the day** |

**TUESDAY, DECEMBER 11, 2001**

*OPEN SESSION*

| | |
|---|---|
| 8:00 a.m. | **Breakfast** |
| 8:30 | **Welcome and Introductions**—Bob Paine, *Chair*, and Susan Roberts, *Study Director* |
| 8:45 | **Lee Alverson,** Natural Resource Consultants, Inc.<br>*Setting the Scene: Trends in Fisheries, Fish Stocks, and Sea Lion Populations* |
| 9:15 | **Russ Andrews,** University of British Columbia<br>*Foraging Ecology of Steller Sea Lions* |
| 9:45 | **Discussion** |
| 10:15 | **Break** |
| 10:30 | **Kate Wynne,** Sea Grant Marine Advisory Program, University of Alaska<br>*Prey Use by Steller Sea Lions Near Kodiak: Recent Findings of the Gulf Apex Predator-Prey Study* |
| 11:00 | **Terrie Williams,** University of California, Santa Cruz<br>*Predator-Prey Energetics for the Steller Sea Lions* |
| 11:30 | **Discussion** |
| 12:00 p.m. | **Lunch** |
| 1:00 | **Tom Loughlin and Anne York,** National Marine Mammal Laboratory, Alaska Fisheries Science Center, National Marine Fisheries Service<br>*Steller Sea Lion and Northern Fur Seal Status and Trends* |
| 1:45 | **Tom Gelatt**, Alaska Department of Fish & Game<br>*Alaska Fish and Game Steller Sea Lion Research, Questions,* |

| | |
|---|---|
| | *Answers, and More Questions* |
| 2:15 | **Discussion** |
| 2:45 | **Break** |
| 3:00 | **Gary Thomas**, Prince William Sound Science Center<br>*Steller Sea Lions and Herring* |
| 3:30 | **Tim Ragen,** Scientific Program Director, Marine Mammal Commission<br>*The Scientific Analysis of Competition Between Alaska Groundfish Fisheries and the Steller Sea Lion* |
| 4:00 | **Discussion** |
| 5:30 | **Meeting adjourns for the day** |

# Appendix G

# National Research Council Project Oversight Boards

## OCEAN STUDIES BOARD

**NANCY RABALAIS** (*Chair*), Louisiana Universities Marine Consortium, Chauvin
**ARTHUR BAGGEROER**, Massachusetts Institute of Technology, Cambridge
**JAMES COLEMAN**, Louisiana State University, Baton Rouge
**LARRY CROWDER**, Duke University, Beaufort, North Carolina
**G. BRENT DALRYMPLE**, Oregon State University, Corvalis
**RICHARD B. DERISO**, Scripps Institution of Oceanography, La Jolla, California
**EARL DOYLE**, Shell Oil (Retired), Sugar Land, Texas
**ROBERT DUCE**, Texas A&M University, College Station
**WAYNE R. GEYER**, Woods Hole Oceanographic Institution, Woods Hole, Massachusetts
**D. JAY GRIMES**, University of Southern Mississippi, Ocean Springs
**MIRIAM KASTNER**, Scripps Institution of Oceanography, La Jolla, California
**CINDY LEE**, State University of New York at Stony Brook
**RALPH S. LEWIS**, Connecticut Geological Survey, Hartford
**BONNIE MCCAY**, Rutgers University, New Brunswick, New Jersey
**JULIAN P. McCREARY, Jr.**, University of Hawaii, Honolulu

**JACQUELINE MICHEL**, Research Planning, Inc., Columbus, South Carolina
**RAM MOHAN**, Gahagan & Bryant Associates, Inc., Baltimore, Maryland
**SCOTT NIXON**, University of Rhode Island, Narragansett
**JON SUTINEN**, University of Rhode Island, Kingston
**NANCY TARGETT**, University of Delaware, Lewes
**PAUL TOBIN**, Armed Forces Communications and Electronics Association, Fairfax, Virginia

**Staff**

**MORGAN GOPNIK**, Director
**SUSAN ROBERTS**, Senior Program Officer
**DAN WALKER**, Senior Program Officer
**JOANNE BINTZ**, Program Officer
**JENNIFER MERRILL**, Program Officer
**TERRY SCHAEFER**, Program Officer
**ROBIN MORRIS**, Financial Officer
**JOHN DANDELSKI**, Research Associate
**SHIREL SMITH**, Administrative Associate
**JODI BACHIM**, Senior Project Assistant
**NANCY CAPUTO**, Senior Project Assistant
**DENISE GREENE**, Senior Project Assistant
**SARAH CAPOTE**, Project Assistant
**BYRON MASON**, Project Assistant
**JULIE PULLEY**, Project Assistant

## POLAR RESEARCH BOARD